Intelligence of Things: AI-IoT Based Critical-Applications and Innovations

Fadi Al-Turjman • Anand Nayyar
Ajantha Devi • Piyush Kumar Shukla
Editors

Intelligence of Things: AI-IoT Based Critical-Applications and Innovations

Editors
Fadi Al-Turjman
Artificial Intelligence Department
Near East University
Mersin 10, Turkey

Ajantha Devi
AP3 Solutions
Chennai, India

Anand Nayyar
Duy Tan University
Graduate School, Faculty of Information Technology
Da Nang, Vietnam

Piyush Kumar Shukla
Rajiv Gandhi Technical University
Bhopal, India

ISBN 978-3-030-82802-8   ISBN 978-3-030-82800-4   (eBook)
https://doi.org/10.1007/978-3-030-82800-4

© The Editor(s) (if applicable) and The Author(s), under exclusive license to Springer Nature Switzerland AG 2021

This work is subject to copyright. All rights are solely and exclusively licensed by the Publisher, whether the whole or part of the material is concerned, specifically the rights of translation, reprinting, reuse of illustrations, recitation, broadcasting, reproduction on microfilms or in any other physical way, and transmission or information storage and retrieval, electronic adaptation, computer software, or by similar or dissimilar methodology now known or hereafter developed.

The use of general descriptive names, registered names, trademarks, service marks, etc. in this publication does not imply, even in the absence of a specific statement, that such names are exempt from the relevant protective laws and regulations and therefore free for general use.

The publisher, the authors, and the editors are safe to assume that the advice and information in this book are believed to be true and accurate at the date of publication. Neither the publisher nor the authors or the editors give a warranty, expressed or implied, with respect to the material contained herein or for any errors or omissions that may have been made. The publisher remains neutral with regard to jurisdictional claims in published maps and institutional affiliations.

This Springer imprint is published by the registered company Springer Nature Switzerland AG
The registered company address is: Gewerbestrasse 11, 6330 Cham, Switzerland

Preface

Artificial Intelligence (AI) and the Internet of Things (IoT) are two incredibly popular acronyms that have newly emerged as a disruptive force within the technological world. In recent years, AI and IoT have revolutionized how the world interacts. Now, this advancement has progressed even further with a new innovative and exciting combination called Artificial Intelligence of Things (AIoT).

The term IoT (Internet of Things) describes the digital network of physical objects (things) in a structure similar to that of the internet. "Things" can be individual devices, but also buildings, living beings, goods, machines, or even complete facilities. The IoT approach collects data from these individual "things" and transfers it to a central storage location either on-premise or in the cloud. This data can then be shared with others, linked up, processed, and used accordingly. An extended approach sees the triggering of event-based actions (e.g., automatic redelivery). If data is stored over a longer period of time, it is possible to not only react more quickly but also plan better.

IoT reaches whole new heights when used in connection with artificial intelligence (AI). When using AI-based applications, any added value comes from the data gathered, which conversely means that every IoT project needs to consider future links with artificial intelligence, because collecting data is not just about generating it, but about using it to its full advantage.

Together, IoT and AI can gather data from a range of different sources (sensors, cameras, microphones), before transforming, correlating, and networking it. Data mining can then be leveraged to create forecasts, gain insights, and trigger actions using workflows you define. In other words, information becomes actions and this is AIoT.

In the book *Intelligence of Things: AI-IoT Based Critical-Applications and Innovations*, we have made a strong attempt to cover the fusion of technologies surrounding AI-IoT, real-time applications, and many more.

The objective of this book is to enlighten various technologies in terms of Artificial Intelligence, Internet of Things, and other converging technologies.

The book comprises 9 chapters, and all the chapters cover the concepts regarding AI-IoT to the best possible manner. Chapter 1 titled "Applications of Internet of

Things (IoT) in Green Computing" gives an introduction to the IoT ecosystem and explains the concept of Green Computing and information and communication technologies (ICT). The chapter emphasizes on many focused areas of Green Computing such as efficient utilization of resources, the establishment of data centers, proper recycling of hazardous materials, and minimizing the effect of the greenhouse. In addition, the chapter focuses on various factors that are useful for crucial planning and decisions making to incorporate significant improvements in the development of IoT Green Computing devices and explores how different IoTGC technologies that are based on virtualization can reduce the demand for hardware resources and power consumption to perform various operations. Chapter 2 titled "Vehicular Intelligence System: Time-Based Vehicle Next Location Prediction in Software-Defined Internet of Vehicles (SDN-IOV) for the Smart Cities" proposes ERS-SDN-IOV approaches used to estimate the shortest and more stable route among the Vehicles in Smart Cities to predict the vehicle future location. The proposed model is tested using Mathematical and Network simulations on Network Simulator 3 and SUMO, and ERS-SDN-IOV shows a less delivery transmission time compared to the AODV, GPSR, and SDGR. The proposed model has 95% delivery ratio and 50 % lesser transmission time as compared to mention existing protocols. Chapter 3 titled "An Enhanced Cloud-IoMT-based and Machine Learning for Effective COVID-19 Diagnosis System" proposes real-time diagnosis system to combat the spread of COVID-19 outbreak. A cloud-IoMT-based framework was developed to collect real-time data for early diagnosis of patients in real time. Four distinct machine learning algorithms Extra Trees, Random Forest (RF), XGBoost, and Light Gradient Boosting Machine (LGBM) are used for quick and better identification of potential COVID-19 cases. The dataset used contains COVID-19 symptoms and selects the relevant symptoms of the diagnosis of a suspected person. The results show that LGBM performs better with an accuracy of 91%, and the least of all the four algorithms shown an accuracy of 75%. Chapter 4 titled "AIIoT for Development of Test Standards for Agricultural Technology" provides information and facts regarding the role of AI and IoT in testing and evaluating agricultural technologies. Chapter 5 titled "Study and Analysis of 5G Enabling Technologies, Their Feasibility and the Development of the Internet of Things" highlights the importance of 5G network in IoT and even elaborates security and protection of 5G networks cum Advantages, feasibilities, and compatibility between 5G and IoT. Chapter 6 titled "Automated Methods for the Detection of Green Land in Satellite Images" proposed a new hybridized U-NET model using ResNet for semantic segmentation of green lands in satellite images. The proposed model, i.e., Skip-U-NET model, is tested and analyzed on DSTL satellite image dataset, and performance comparison of the proposed model is done as compared to existing models. Chapter 7 titled "Artificial Cyber Espionage Based Protection of Technological Enabled Automated Cities Infrastructure by Dark Web Cyber Offender" discusses all cyber security threats penetrating the future smart cities and proposed solutions to counterfeit all such types of attacks. Chapter 8 titled "Role of Artificial Intelligence and IoT in Next Generation Education System" explores the possibilities of incorporating AI and IoT in the future education system in India and the various practices that are being followed by education industries/sectors to make it a reality. Chapter 9 titled "Social Media Data Analysis: Rough Set Theory Based Innovative Approach" proposes innovative approach based on rough set in social media data analysis.

The book will be a strong reference to students, research and development engineers, and industry professionals to understand the foundations of the advanced concept of AI-IoT.

Mersin 10, Turkey Fadi Al-Turjman
Da Nang, Vietnam Anand Nayyar
Chennai, India Ajantha Devi
Bhopal, India Piyush Kumar Shukla

Contents

Applications of Internet of Things (IoT) in Green Computing 1
Ankit Garg and Anuj Kumar Singh

Vehicular Intelligence System: Time-Based Vehicle Next Location
Prediction in Software-Defined Internet of Vehicles (SDN-IOV)
for the Smart Cities 35
Preeti Rani, Naziya Hussain, Rais Abdul Hamid Khan,
Yogesh Sharma, and Piyush Kumar Shukla

An Enhanced Cloud-IoMT-based and Machine Learning
for Effective COVID-19 Diagnosis System 55
Joseph Bamidele Awotunde, Sunday Adeola Ajagbe,
Ifedotun Roseline Idowu, and Juliana Ngozi Ndunagu

AIIoT for Development of Test Standards for Agricultural
Technology 77
Puneet Kumar Aggarwal, Parita Jain, Poorvi Chaudhary,
Riya Garg, Kshirja Makar, and Jaya Mehta

Study and Analysis of 5G Enabling Technologies, Their Feasibility
and the Development of the Internet of Things 101
Rubaid Ashfaq

Automated Methods for the Detection of Green Land in Satellite
Images 145
Raju Pal, Subash Yadav, Aarti, Pushpendra Kumar Rajput,
and Anand Nayyar

Artificial Cyber Espionage Based Protection of Technological
Enabled Automated Cities Infrastructure by Dark Web
Cyber Offender 167
Romil Rawat, Vinod Mahor, Sachin Chirgaiya, and Bhagwati Garg

Role of Artificial Intelligence and IoT in Next Generation Education System 189
Kiran Ahuja and Indu Bala

Social Media Data Analysis: Rough Set Theory Based Innovative Approach 209
K. Anitha

Index 227

About the Authors

Fadi Al-Turjman received his PhD in computer science from Queen's University, Canada, in 2011. He is the *associate dean for research* and the *founding director* of the International Research Center for AI and IoT at Near East University, Nicosia, Cyprus. Prof. Al-Turjman is the head of Artificial Intelligence Engineering Dept., and a leading authority in the areas of smart/intelligent IoT systems, wireless, and mobile networks' architectures, protocols, deployments, and performance evaluation in Artificial Intelligence of Things (AIoT). His publication history spans over 400 SCI/E publications, in addition to numerous keynotes and plenary talks at flagship venues. He has authored and edited more than 40 books about cognition, security, and wireless sensor networks' deployments in smart IoT environments, which have been published by well-reputed publishers such as Taylor and Francis, Elsevier, IET, and Springer. He has received several recognitions and best papers awards at top international conferences. He also received the prestigious *Best Research Paper Award* from Elsevier *Computer Communications* journal for the period 2015–2018, in addition to the *Top Researcher Award* for 2018 at Antalya Bilim University, Turkey. Prof. Al-Turjman has led a number of international symposia and workshops in flagship communication society conferences. Currently, he serves as book series editor and the lead guest/associate editor for several top-tier journals, including the *IEEE Communications Surveys and Tutorials* (IF 23.9) and the Elsevier *Sustainable Cities and Society* (IF 7.8), in addition to organizing international conferences and symposiums on the most up to date research topics in AI and IoT.

Anand Nayyar received his PhD (Computer Science) from Desh Bhagat University in 2017 in the area of Wireless Sensor Networks and Swarm Intelligence. He is currently working in the School of Computer Science-Duy Tan University, Da Nang, Vietnam, as Assistant Professor, Scientist, Vice-Chairman (Research), and Director—IoT and Intelligent Systems Lab. He is a Certified Professional with 75+ professional certificates from CISCO, Microsoft, Oracle, Google, Beingcert, EXIN, GAQM, Cyberoam, and many more. He has published more than 125+ research papers in various high-quality ISI-SCI/SCIE/SSCI impact factor journals cum Scopus journals; 50+ papers in international conferences indexed with Springer, IEEE Xplore, and ACM Digital Library; 40+ book chapters in various SCOPUS and WEB OF SCIENCE indexed books with Springer, CRC Press, Elsevier; and many more with citations: 4200+, H-Index: 35, and I-Index: 113. He is a member of more than 50+ associations as senior and life member including IEEE, ACM. He has authored/coauthored cum edited 30+ books on computer science. He is associated with more than 500+ international conferences as program committee/chair/advisory board/review board member. He has 11 Australian patents, 5 Indian patents, and 1 Indian Copyright to his credit in the area of wireless communications, artificial intelligence, cloud computing, IoT, and image processing. He was awarded 32+ awards for Teaching and Research—Young Scientist, Best Scientist, Young Researcher Award, Outstanding Researcher Award, Excellence in Teaching, and many more. He is acting as Associate Editor for *Wireless Networks* (Springer), *Computer Communications* (Elsevier), *International Journal of Sensor Networks* (IJSNET) (Inderscience), *Frontiers in Computer Science*, *PeerJ Computer Science*, *Human Centric Computing and Information Sciences* (HCIS), *IET-Quantum Communications*, *IET Wireless Sensor Systems*, *IET Networks*, *IJDST*, *IJISP*, *IJCINI*, and *IJGC*. He is acting as Editor-in-Chief of IGI-Global, USA Journal titled *International Journal of Smart Vehicles and Smart Transportation (IJSVST)*. He has reviewed more than 1400+ articles for various Web of Science indexed journals. He is currently conducting research in the area of wireless sensor networks, IoT, swarm intelligence, cloud computing, artificial intelligence, drones, blockchain, cyber security, network simulation, and wireless communications.

About the Authors

Ajantha Devi is working as a Research Head in AP3 Solutions, Chennai, Tamil Nadu, India. She received her PhD from the University of Madras in 2015. She has worked as Project Fellow under UGC Major Research Project. She is a Senior IEEE Member. She has been certified as "Microsoft Certified Application Developer" (MCAD) and "Microsoft Certified Technical Specialist" (MCTS) from Microsoft Corp. She has more than 35 papers in international journals and conference proceedings to her credit. She has written, coauthored, and edited a number of books in the field of computer science with international and national publishers like Elsevier and Springer. She served as a member of the Program Committee/Technical Committee/Chair/Review Board for a variety of international conferences. She has five Australian patents and one Indian patent to her credit in the area of artificial intelligence, image processing, and medical imaging. Her work in image processing, signal processing, pattern matching, and natural language processing is based on artificial intelligence, machine learning, and deep learning techniques. She has won many best paper presentation awards as well as a few research-oriented international awards.

Piyush Kumar Shukla (PDF, PhD, SMIEE, LMISTE) is Associate Professor in Computer Science and Engineering Department, University Institute of Technology, Rajiv Gandhi Proudyogiki Vishwavidyalaya (Technological University of Madhya Pradesh). He has completed his Post Doctorate Fellowship (PDF) recently in March 2020 under "**Information Security Education and Awareness Project Phase II" funded** by the Ministry of Electronics and Information Technology (MeitY), from SVNIT Surat, Gujarat, India, from the Department of Computer Engineering. He holds a PhD in CSE, MTech in CSE, BE in ECE from RGPV, Bhopal, MP, India.

Dr. Shukla has published more than 150 research papers, book chapters, and books at the national/international level in *IEEE/ACM Transactions*, Elsevier, and Springer. He is currently editing 07 books with Springer, IGI-Global, Taylor & Francis, Wiley-SP, and AAP. He is an active reviewer and editorial member of more than ten reputed international journals in his research areas, such as IEEE Transactions, Elsevier journals, and Springer journals. He has teaching and research experience of 15 years. He has been Assistant Registrar, Academic (PhD), Co-Coordinator of Intellectual Property Right Cell (RGPV), and Co-Coordinator of AICTE at RGPV. Dr. Shukla has been delivering several expert lectures in online and offline mode. He is the recipient of many awards and recognitions like Selected for Best Researcher Award-2020, 2nd International Research Awards on Science, Health and Engineering by International Research Awards Science father, India. Dr. Piyush has delivered more than 25 expert lectures and chaired many technical sessions in international conferences worldwide, such as in the USA. He is also a member of many professional bodies like **IEEE (Senior Member)**, and **ISTE (Life Member)**. He has Project Grants under progress: 01, Principal Investigator (PI), Titled **"Precision Agriculture: Smart Farming with IoT and Drone for Increasing Productivity of Crops in India"** worth Rs. 3.0 Lakh under research grant funded by TEQIP-III, RGPV, Bhopal. He has applied for other project grants in SERB, DST, MPCST scheme, and several other Govt. of India. His research interest includes machine intelligence, medical image processing, and white-box cryptography. He has supervised 07 PhD scholars and 50 M. Tech dissertations to date. He has served the department in PG In-charge/Coordinator MTech, DDIPG programs, Department Coordinator NBA, NAAC, and NIRF & as HoD several times.

Applications of Internet of Things (IoT) in Green Computing

Ankit Garg and Anuj Kumar Singh

1 Introduction

IoT is the combination of two essential terms "Internet" and "Things." The world has now become smart with the development and improvement of science and technology. With the development of various technologies, the users of IoT networks are collaborating themselves using various IoT-enabled devices. The use of various sensors, controllers, actuators, and other miscellaneous objects enables people to link with the digital world for sharing information through the Internet of Things (IoT). The sharing of information in this way provides intellectual perceptions into human beings [1–5]. All the IoT devices such as RFID tags and their inbuilt sensors consume more energy to process the information they exchange over the network [6, 7]. Big data and cloud computing are interlinked with the IoT to investigate the behavior of IoT enables devices. Nowadays organizations are trying to adapt to the environment in which the demands of IoT users are increasing exponentially. According to the available literature, by 2025, the IoT nodes can be inbuilt in everything such as washing machines, food packages, documents, etc. For the establishment and betterment of the greener and smart world, IoT devices should be easily acceptable, eco-friendly, easily disposable, energy-efficient, and economically viable [8, 10, 11]. IoT devices are making strong utilization of diverse cloud platforms. These devices store and transmit the information of sensed data to cloud providers like Thingspeak.com to facilitate the users for real-time monitoring of devices [12]. The concept of green computing explores various prominent areas such as designing and manufacturing of products, disposing of electronic devices, and suggests the possible ways to minimize the hazardous impact on the environment.

A. Garg (✉) · A. K. Singh
Amity University, Gurgaon, India
e-mail: agarg1@ggn.amity.edu; aksingh@ggn.amity.edu

The methods of green computing promote the utilization of waste materials (recycling) and develop energy-efficient hardware and software systems. Green computing provides a sense of social responsibility to the people in the public sector to minimize the financial requirements of the nation. The concept of "Green Computing" originated by the US Environmental Protection Agency (EPA) in 1992 with the launch of a voluntary labeling program, i.e. Energy Star, to promote energy efficiency in computer hardware.

Nowadays, various agencies are promoting green computing by providing certification like Energy Start Label to computer systems. Similarly, the TCO certification program that is launch by the Swedish organization TCP development promotes the consumption of low emission components. Researchers have been explored various areas of research in the field of IoT-based green computing such as the design and development of IoT enables devices, managing operations, and their maintenance. This field of research fascinates the researchers to explore the challenges and issues related to change of climate, the crisis of required energy, and other environmental issues [9]. Nayyar et al. [13] provide in-depth view of Internet of Nano Things (IoNT), its architecture, application areas, and various challenges to make researchers aware of IoNT standards. IoT ecosystem is used to connect heterogeneous IoT devices, controllers, servers, gateways, and platforms efficiently into a single system. Various communication protocols and interfaces such as low-power Wi-Fi, Near Field Communication (NFC), Zigbee, etc., are used to connect these components to exchange the information. The block diagram in Fig. 1 presents the IoT ecosystem.

Information and Communication Technologies (ICT) have now been automated to make various routine tasks easier. Computers and the Internet play a significant role to improve various processes of the ICT sector in the real world. Billions of users are getting a response to their queries using the "Client-Server" model. A huge amount of information can be distributed worldwide that store on the servers of social websites and networks. In addition to this, with the development of a machine to machine communication various systems over the network are used to provide information and services to other machines. Till date, more than 50 billion IoT devices have been connected to the IoT networks. Researchers have been predicted that in the next coming years, this number can be increased up to 3 trillion. Green ICT is required to maintain green IT to create a healthy environment with the optional use of the Internet and other IT resources. All IoT-enabled devices are required ample amounts of energy for their operations and put a large encumbrance on the electrical grid and greenhouse gases. The regular change in the pattern of climate and the accumulation of greenhouse gases may be the cause of drought and flood in some areas. The exponential growth in the temperature of the earth may be responsible for the severe environmental glitches. From various studies carried out by researchers present that the number of weather-related disasters is increasing every year. To overcome this situation we need to focus on the growth of global emission of greenhouse gases and their preservation. Unnecessary consumption of energy emitted from natural resources such as coal and oil must be suppressed which can be highly responsible for the drastic change in the environment. The deduction in

Fig. 1 Block diagram of IoT ecosystem

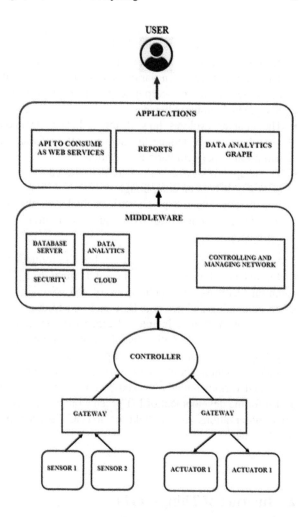

unnecessary electric power consumption can lay hold the acute impact of global warming. To minimize the impact of CO_2 in the environment a concrete road map must be designed in which existing technologies of IoT can be utilized. The manufactures should adopt the advanced design for manufacturing products that require less energy and other natural resources. The exponential growth of the population and the integration of IoT devices over the network are a never-ending process. To change the utilization and buying behavior of the users, some green policies should be formulated. These policies and regulations motivate users to use only green products. Besides this, the traditional methods should be exterminated and new trends must be incorporated to build a green environment.

The major contributions of the chapter can be summarized as follows:

- The chapter gives an introduction to the IoT ecosystem, explains the concept of green computing and information and communication technologies (ICT).

- The chapter emphasis on many focused areas of green computing such as efficient utilization of resources, the establishment of data centers, proper recycling of hazardous materials, and minimizing the effect of the greenhouse, etc.
- The chapter focuses on various factors that are useful for crucial planning and decisions making to incorporate significant improvements in the development of IoT green computing devices.
- The chapter explores how different IoTGC technologies that are based on virtualization can reduce the demand for hardware resources and power consumption to perform various operations.
- The chapter reviewed various aspects related to awareness to improve the efficiency of IoT green computing.
- Finally, the chapter explores various challenges that are needed to be focused to recycle the products to attain successful environmental-friendly IoT green computing.

In this chapter, Sect. 1 provides a brief introduction to the Internet of Things (IoT), IoT-based green computing, and green ICT. Section 2 explains various common elements and applications of IoT. Section 3 presents an overview of green computing and green IoT. In addition to this, it provides knowledge to the readers about various principles of green IoT. Section 4 explains various applications of IoT green computing based on five different categories such as hardware, recycling, awareness, software, and policy. The various subsections of this section provide the details about how real-time software and hardware, awareness of the users, and various initiatives taken by the government and private agencies can be useful for the development of IoT green Computing environment. Finally, based on some observations in the field of IoTGC, Section 5 presents discussion and conclusions.

2 Internet of Things (IoT)

IoT is widely used to connect billions of physical devices through sensors and other wireless technologies to the Internet. It provides facilities for the use of IoT systems to exchange massive information for effective decision making. It enables security, data management, resource management, and other useful operations to provide better service to the IoT users. According to the available literature it has been suggested that by 2025, billions of IoT-enabled devices may be connected to the IoT networks. Various key segments of IoT such as hardware, middleware, Internet, and presentation are used for gathering data to provide the best services to the users.

Figure 2 presents various common elements of IoT.

- *Identification:* Through this key element a unique identification is provided to the different IoT devices such as Ubiquitous Code (Ucode), object identifier (OID), Radio-Frequency Identifier (RFID), Universal Unique Identifier (UUID),

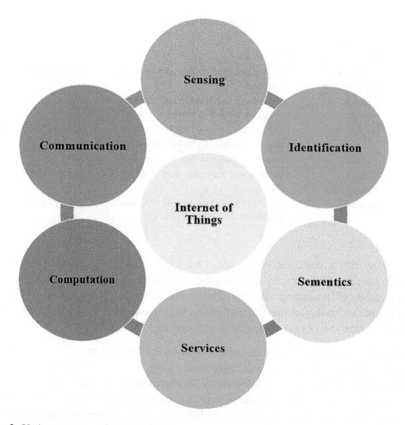

Fig. 2 Various common elements of IoT

Automatic Identification and Data Capture (AIDC), and International Mobile Equipment Identity (IMEI), etc.
- *Sensing:* To make the system smarter the sensors gather and distribute the data among various connected devices of the network. These sensors are used to sense useful information such as temperature, acceleration, pressure, chemical reactions, etc.
- *Communication:* In wireless communication, many communication protocols are widely used such as Internet Protocol Version 6 (IPv6), Bluetooth, Wi-Fi, over Low-power Wireless Personal Area Networks (6LoWPAN), Z-Wave, and ultra-wide bandwidth (UWB), etc.
- *Computation:* Various common hardware platforms such as Raspberry Pi, Arduino (Genuino), ESP 8266, Intel Edison, Intel Galileo, etc. and software platforms such as Axeda, Nimbits, SensorCloud, etc., are used to carry out various computational tasks.
- *Services:* A large number of IoT applications require a particular IoT service. These services can be of four types such as information aggregation services, identity-related services, collaborative-aware services, and ubiquitous services.

Table 1 Segmentation of various IoT application domains into different categories [16–19]

S. no.	Application domain	Description
1	Application Development	Development and deployment of the IoT-enabled products by the companies
2	Network Management	Establishing and maintaining communication among various IoT entities on the IoT networks
3	Data Analytics	Data produced by various IoT devices can be analyzed based on various classification and clustering techniques
4	Visualization	Presenting analyzed data in an understandable form
5	Data Management	Handling and managing data efficiently using various existing technologies
6	Research	Use of IoT in different fields of research such as green IoT for a sustainable environment [14, 15]

- *Semantics:* Various common semantics are widely used to extract the information into a meaningful form by adopting the steps such as identification and proper utilization of resources, data modeling, and analysis. The most widely used semantic technologies are Efficient XML Interchange (EXI), Resource Description Framework (RDF), Web Ontology Language (WOL), etc.

The architecture of IoT can be subdivided into various layers such as perception (Layer-1), network (Layer-2), and application (Layer-3). The architecture of IoT is further extended by adding middleware and business layer. The middleware layer is also known as processing layer that processes and analyzes the massive collection of data supplied by the network layer. Establishing connection and managing various services are the major responsibilities of this layer. The business layer deals with the well-constructed business models for decision making. This layer plays a major role in various business strategies, user privacy, and other applications. Different application domains of IoT and their description are shown in Table 1.

Nowadays the concept of IoT is widely used in various significant areas such as the development of smart cities, smart transportation system, smart health care systems, home automation systems, smart agriculture systems, etc. [20–22]. Gathered and analyzed data related to various applications of IoT can be used in various commercial applications. Various IoT devices are widely used to gather useful data for the establishment of IoT-based smart cities. The smart cities use the data to provide better services to the users and also improve the infrastructure of the smart cities.

IoT-enabled smart health care systems provide better health care to the patients by maintaining and analyzing their regular health records. These systems are used to check the availability of medical resources and error-free communication between physicians and patients. The existence of all these facilities in the IoT-based health care system provides the trust and reliability of IoT systems among the users [23–25]. Figure 3 presents various IoT applications.

Applications of Internet of Things (IoT) in Green Computing

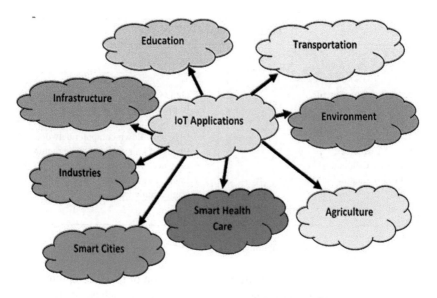

Fig. 3 Various applications of IoT

For IoT-based transport systems various efficient framework has been suggested by the researches. These frameworks exploit the collected data by different IoT sensors to process the information among the IoT-enabled devices. These frameworks provide the facilities of better communication, coordination, safety, and control in the transportation system. IoT-based smart vehicle monitoring system can be used to detect the accidents at a very early stage so that any miss happening can be stopped. IoT-enabled devices can be used for the construction of the smart home and other infrastructure systems. IoT-based technologies are best suited for various service-oriented industries, IoT-based agriculture, environment and energy management, security and monitoring systems, and various other similar kinds of areas.

3 Green Computing and Green IoT

The demand for energy to operate various IoT-enabled devices is increasing day by day. Different manufacturing companies are trying to develop efficient technologies to reduce the requirement of energy. Researchers have been suggested that by 2030 the number of IoT devices can be reached up to 100 billion. In the same proportion, the level of carbon dioxide is exponentially increasing due to the heavy use of IoT devices. The study carried out by the researchers reveals the fact that by 2020 approximately 344 tons of carbon dioxide could be generated from cell phones. The exponential growth of CO_2 in the environment can be harmful to the health of the human being. Therefore, to maintain a sustainable environment globally it is

GREEN COMPUTING	
PROS	**CONS**
1. Waste Management	1. Quite Costly
2. Proper Utilization of resources	2. Rapid Change in Technology
3. Optimization	3. Barely Available
4. Risk Management	4. May slow down computer network
5. Cost Effective	5. Plenty of knowledge may be required
6. Energy Efficient	6. Low acceptance rate
7. Environmental Sustainability	7. Lack of awareness
8. Effective Branding	8. High maintenance cost

Fig. 4 Pros and cons of Green Computing

required to take some major steps towards the inclusion of green computing in IoT systems. There are many focused areas of green computing such as efficient utilization of resources, the establishment of the data centers, proper recycling of hazardous materials, minimizing the effect of the greenhouse, etc. [14, 15]. Various efficient technologies of green computing can be incorporated to minimize the adverse effect of greenhouse gases and their impact on the global environment. According to the literature and case studies, the hazardous impacts of greenhouse gases have been reduced after adopting the technologies of green computing [18, 20, 26–28]. Nowadays the demand for energy utilization is exponentially increasing to operate various IoT devices. To fulfill the energy consumption requirements companies and their subscribers are adopting green IoT. Manufactures and other service providers are using various recent technologies to save energy in smart homes and buildings. An ample amount of energy can be conserved technologies related to air-conditioning and ventilation and are implemented efficiently.

Various techniques have been suggested by the researchers to save energy in Wireless Sensor Network [29]. Perera et al. [30] suggested an orchestration agent (OA) which is used for the evaluation of resource consumption of the servers and also ensures reliability. Figure 4 shows the pros and cons of green computing in the current scenario.

Murugesan et al. [31] have been proposed an efficient C-MOSDEN model to reduce the requirement of power consumption. The suggested scheduling algorithms in the proposed model are used to accumulate the energy in IoT sensors. In smart health care systems, IoT devices are widely used to store the real-time data of the patient [32]. Kim et al. [33] have been suggested an efficient algorithm that manages the consumption of energy in cloud storage as well as in the exchange of the data.

3.1 Principles of Green IoT

In today's scenario, it can be observed that conventional energy resources are obsoleting day by day. Various IoT-enabled devices are widely used among IoT users that require significant energy consumption. With the exponential growth in power consumption through IoT devices, the concept of green IoT plays a major role in ICT research. Researchers have been suggested the below-mentioned principles of green IoT to gain the benefits from it.

- In the IoT system different nodes are required to be placed diligently and to conserve energy-efficient routing techniques can be adopted.
- IoT-enabled devices are required to gather the most essential data and wipe out the irrelevant data which require a high amount of energy consumption.
- Deployment of passive and active sensors on the network can significantly reduce the amount of energy consumption.
- Effective decisions and policies are required to be developed to reduce the exponential growth of energy utilization by the smart transport system, smart health care system, etc.
- Cost-effective trade-offs are required to be selected based on cost, communication, and processing.

4 IoT in Green Computing

The communication process of IoT to exchange the information must be energy efficient. In exchange of the data over network different IoT devices, sensors must consume less energy to perform various operations. Various applications of the IoT green computing can be used in various areas such as the development of smartphones [16], efficient utilization of data storage [18], development of IoT-based ecosystem [19], development of an energy-efficient system to exchange the data [20], smart health management and agriculture system [9], and in the establishment of Green Campus [21]. Various application areas of IoT in green computing can be divided into different subcategories as depicted in Fig. 5.

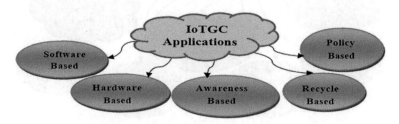

Fig. 5 Applications of IoT Green Computing

4.1 Applications of IoT Green Computing Based on Policies

Based on real-time data the crucial planning and decisions can be taken at different levels for the development of IoT green computing devices. The effective decision-making process can supply various inputs to give an impact on the environment. Proper management of data, better planning for energy consumption, and an industrial automation system can be helpful for the establishment of effective IoT green computing. In a home automation system, proper data gathering and its processing can minimize the significant amount of energy depletion [22–25]. Figure 6 presents various applications of policy-based IoT green computing.

4.1.1 IoT-Based Green Campus

With the exponential growth of population, the limited number of resources are depleting rapidly. For example, in 1995, the atmospheric CO_2 concentration was less than 320 ppm. After the industrial revolution, it has been significantly increased by up to 25% [18]. It is an indicator of a deteriorated situation that presents a contaminated and unhealthy environment around us. The other problems such as a sudden difference in temperature, soil degradation, and acid rain cannot be ignored completely. Government and other organizations have to take some initiatives to minimize the carbon emissions in the environment. For a better civilization government have to develop some energy proficient systems. Proper promotion of policies and awareness of the green world among the human being can play a crucial role in

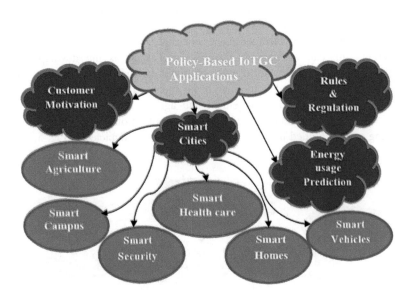

Fig. 6 Applications of policy-based IoT Green Computing

the development of the green campus or residence. These smart campuses must be equipped with an automated monitoring and control mechanism to manage several facilities economically. The purpose of a smart campus is to provide the facilities that require a minimum amount of energy [20]. The advanced technologies in the smart campuses effectively control different facilities such as air conditioners, computer laboratories, etc., and establish a planned IoTGC system that can save electricity.

4.1.2 Green Agriculture and Green Health Care

In conventional health care and agriculture systems, IoT green computing can play a major role. IoT green computing-based health care systems have some applications such as smart health monitoring systems, handling a massive amount of data using data mining and artificial intelligence techniques. Similarly, a smart agriculture system enables farmers to increase crop yields, and also aware farmers about the efficient technology of irrigation. Efficient farming with modern IoT-enabled equipment can reduce the cost of farming [20, 21]. The integrated IoTGC model with these systems can improve versatility, intelligence, and interoperability. The administration cost in both types of smart system can be minimized due to the use of IoT-enabled device. Such tools share information over the Internet through different modules of the network, thus greatly simplify and reduce the tasks of administration. For example, in case of an emergency, the automatic alarming system can provide information about the nearest reachable hospital. Similarly, IoTGC based smart farming system can improve the crop yield using several IoT-enabled devices such as moisture sensors, temperature sensors, etc. [9].

4.1.3 Intelligent Automobiles

In IoTGC based smart transportation system various efficient techniques have been investigated by the researchers. Through IoTGC based smart vehicle systems, time and cost can be significantly saved. In these automated systems various IoT- based sensors are used to detect the possibility of an accident in advance. These systems also provide e-notification about traffic and weather conditions via remote servers. Through IoT sensors, the speed of the vehicles can be controlled in heavy traffic and bad weather condition.

The automatic alarming system can estimate the response time and travel time of the drive to generate information about the miss happening to his family. Likewise, the use of GPRS tracing scheme, GSM dial-up connection, and infrared proximity sensors are some of the few examples that can be helpful in the smart vehicle management system [23]. The smart vehicle management mechanism reduces the pollution due to unnecessary blockage of traffic and also save the oil from the excessive acceleration and braking of vehicles [24]. The equipped vehicle monitoring sensors can identify the state of the vehicle, its owner, and can provide the appropriate support by alerting the agencies in uncontrollable conditions on the road [34, 35].

4.1.4 Intelligent Houses

The IoTGC can be used to design the map of intelligent houses that can provide comfort to the end-users. Smart homes provide various facilities to the users so that they can be capable to manage their daily scheduled and plans. Usually, smart houses are equipped with advanced software, sensors, smart monitoring systems, smart alarming systems, etc. [36]. All IoT-enabled devices are interconnected with each other through a gateway to provide the necessary information to the end-user on time. IoTGC emerges with a revolution that the devices and humans must be connected in every space and time to make life easier. Significant innovations in the field of IoTGC transform human life and force manufactures to developed eco-friendly devices. These eco-friendly devices can be helpful to create a clean and healthy climate around the human being. The reduction of energy consumption through IoT-enabled devices can manage the utilization of the limited resource of the world such as water, oil, and other sources of energy.

4.1.5 Intelligent Protection/Security

Due to significant improvements in the World Wide Web the human being and hardware devices are now linked together at every place and time. The smart security system based on IoT can provide a way to disseminate sensitive information among various private and government agencies. The use of IoT-based gateways can prevent the leakage of sensitive information over the network. Various password protection mechanisms, encryption, and decryption techniques, and efficient routing algorithms can be used to secure the information. The smart protection system avoids unnecessary threats that can theft information related to the public and government sectors of the nation [37]. Furthermore, various factors such as integrity, confidentiality, and reliability of the information should be maintained over perception and its sub-layers. The smart access control mechanism should be implemented over the IoT network so that unauthorized access can be restricted. The efficient fault tolerance technique should be adopted to manage the various operations and communication among the devices.

4.2 *IoTGC Applications Based on Software*

In Fig. 7 various applications based on the IoTGC are presented. Various applications of IoTGC based on software are explained in the subsequent sections.

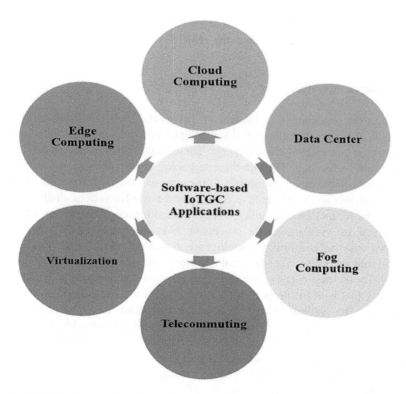

Fig. 7 IoTGC Applications based on software

4.2.1 Cloud Computing

Cloud computing provides services to the user to access servers, databases, and various other application services over the Internet. The five main characteristics of cloud computing are (1) measured services, (2) quick provision of the resources, (3) resource pooling, (4) broad network access, (5) on demand self-service. In distributed and parallel networks various applications and hardware can be utilized as essential services.

Nowadays industries and users are trying to shift their attention toward managing various business activities through virtualization. Through this cloud computing method, a virtual version of storage, hardware, and software can be created. Virtualization of various resources can save the time of many organizations. In a cloud computing environment, various virtualized systems are intertwined together [17, 18].

Virtualization of various resources in cloud computing reduces the emission of carbon in the environment and also minimizes the consumption of energy to a great extent. Online shopping system over commercial websites is a good example of virtualization. The user can purchase various products and demand for the services without being present physically at the supermarket. Through online shopping traffic on the roads can be reduced that results saving of unnecessary consumption of vehicle oil and emission of gas from the vehicles.

4.2.2 Edge Computing

Nowadays the applications of Internet of Things (IoT) are rapidly growing. The conventional cloud computing-based technologies encounter several challenges like high latency, low spectral efficiency, etc. These challenges motivate researchers to develop new technology. Through these technologies various functions of conventional cloud computing can be shifted to the edge devices of the network. Various technologies that are based on edge computing can resolve various challenges that are being faced in cloud computing. In edge computing, edge devices are the smart devices that are mainly used for processing and computation. It provides server resources, data analysis, artificial intelligence to data collection sources, and cyber-physical sources. These cyber-physical resources can be smart sensors and actuators [38, 39]. The developers can create various applications by managing and harnessing the energy that is available in remote locations such as hotels, warehouses, retail stores, etc. The applications based on edge computing can minimize latency, improve privacy, minimize the demand of the network bandwidth, and provide better services even when the network is disrupted. Edge computing brings data processing to the edge of the network, where speed and reliability are critical to a successful customer experience. The key components of edge computing are discussed below and presented in Fig. 8.

Cloud: The clouds are the server over the Internet. These servers contain various applications and databases that are being accessed by users and companies. The concept of cloud facilitates the users to manage their data without managing their physical servers. Without using their systems, users can directly execute their applications on the clouds.

Edges: An edge is a network edge where the devices are used to communicate with the Internet. At the network edge, various edge nodes such as servers, gateways, and other edge devices are used to perform various computations. These edge devices are used to perform various tasks related to the assembly machines in a factory, an intelligent camera inside the house, and automobile companies. Generally, edge devices like edge servers can have multiple cores of computing capacity and 12–16 GB of memory. These servers are used to execute software of industries to exchange services to the clients. Similarly, an edge gateway performs various functions over the network like network termination, firewall protection, etc. The edges in edge computing are generally considered in between the path of data sources and cloud data centers.

Sensors, controllers, and actuators: Sensors in the Internet of Things (IoT) play a major role to collect various information such as information about temperature, moisture, etc. The important data that is being collected by the sensors is first tuned according to the need of the customer or user and transmitted over the network. The tuned data can be used by the user to perform their useful tasks. For an effective

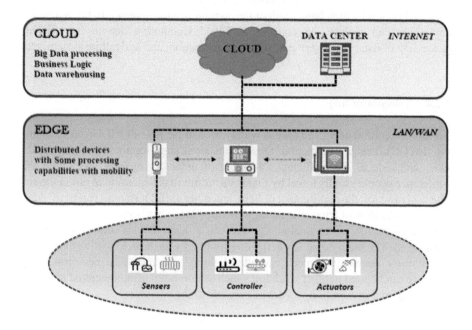

Fig. 8 Key components of edge computing

decision-making process, the controllers figure out how data can be utilized that is being collected by the sensors. The actuator works in opposite direction of a sensor. It receives input in the form of electrical signals and performs the physical action accordingly. An electric motor and hydraulic system can be good examples of actuators.

4.2.3 Fog Computing

The concept of Fog computing improves performance and significantly reduces the amount of information, its processing, and storage requirements. It is a decentralized computing infrastructure in which data, its storage, applications are usually located in between the source of data and clouds. Fog computing integrates various sensors that are utilized at the edge of the network and provides the services to the cloud's core computing structure [40]. Through fog computing quality of service, security in real-time systems and stability of the real-time system can be achieved in a better way as compared to Cloud Computing. In fog computing, useful data can be processed on a data hub, routers, and gateway devices [41, 42].

The concept of fog computing supports the IoTGC when the different sensors produce an excess amount of information and find itself incompetent in its transmission through the cloud. The transmission of the data among various sensors on the cloud requires a high bandwidth of the communication channel. In future the fog computing can replace the concept of Cloud Computing due to its efficient processing of data, better services, smart transportation, and agriculture system, etc.

4.2.4 Telecommuting

Various teleconferencing software is widely used in the field of IoT Green Computing (IoTGC). The development of these applications significantly improves productivity and reduces the inefficient utilization of time. To organize and attend the conference people usually travel by flights that result in the emission of gas in the air. The cost of space, the requirement of light, and other useful resources can be avoided after the utilization of teleconferencing software [23]. It has been reported that using these software 70% of the total energy that was utilized in a US office building has been saved. The main cause of energy consumption was the heavy utilization of air conditioners, heaters in winters, and other lighting resources [43]. Telecommuting software can be used in other work lines such as telecommuting-integrated hoteling, consulting, and distribution of various services to the end-user.

4.2.5 Based on Data Center

Data centers are generally responsible for the consumption of energy. The data centers need to be optimized to save the wastage of energy. Various efficient power scheduling algorithms can be used to minimize unnecessary power consumption by passive sensors. The algorithms supply power only to the active sensors to minimize the requirement of the energy over the network [29]. To manage the data centers efficiently following challenges can be faced.

- Estimation and optimal allocation of the workload judiciously to the various servers.
- Recognition of future ideal servers, and workload reallocation.
- Use of renewable generators.
- An effective utilization of available energy and resources.

4.2.6 Virtualization Based

The IoTGC technology based on virtualization significantly reduces the demand for hardware resources to perform various operations. The usage of Mixed Integer Linear Programming (MILP) in the virtualization process of four-layer architecture

Applications of Internet of Things (IoT) in Green Computing

results in energy preservation up to 35% [8]. The advantages of virtualization-based IoTGC are given below.

- Management of data centers in an efficient way.
- Risk management in various business operations.
- Processing of requests and services in a faster way.
- Improving performance, profitability, responsiveness, and agility.
- Reduction in operational cost and smooth IT operations.
- Redistribution of workload and automation of various operations.

4.3 Hardware-based IoTGC Applications

In the IoTGC system, efficient hardware design plays a crucial role to make the system eco-friendly and efficient in terms of power consumption. Few of the IoTGC hardware modules have been presented in Fig. 9 and discussed in the subsequent sections.

4.3.1 RFID Based

The detection of the radio frequency (RFID) remains crucial in IoT green computing. These are the wireless microchips that identify dissimilar substances automatically. These chips are used to collect data in a cost-effective manner related to various business activities of an organization that are expended in various continents. In the IoTGC network, various functions of the network objects can be easily identified and managed to obtain real-time data. The real-time data stored in the database can be further used in various research and business-related activities [44]. The utilization

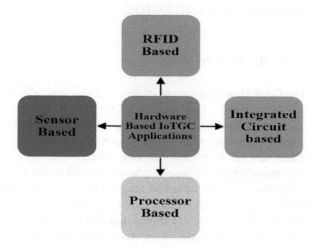

Fig. 9 Hardware-based IoTGC Applications

of RFID in the field of IoT green computing can provide significant improvements in various areas such as industry, health care, online payment system, military projects, and operations, etc. [45]. Anuj et al. suggested elliptic curve signcryption based security protocol for RFID and mutual authentication protocol for smart card [46, 47]. The challenges related to the use of RFID technology in various applications of IoTGC are given below.

- *Collision issues:* Electromagnetic interference among multiple broadcasts can cause problems. Various efficient protocols such as the binary tree, question tree, etc. can be used to resolve the problem of collision. The efficiency of these protocols is however below 50%, hence advanced techniques and protocols are needed to be designed.
- *Privacy and security concerns:* The use of RFID also plays a major role to resolve the traffic analysis problem. In an efficient security mechanism of RFID tags results in DoS attacks, eavesdropping of targets, and can create problems in traffic analysis. Researchers are trying to implement more secure algorithms and protocols that can efficiently manage the RFID system. Efficient protocols can preserve the applicability of RFID systems in the field of IoTGC.

4.3.2 Based on Integrated Circuit

The improper design of integrated circuits that are used in the IoTGC network can create various issues related to power consumption. Integration of multiple sensors in a single chip can significantly minimize the carbon footprint and consumption of energy. Balanced IC technologies based on logic platforms such as IPs, Fin Field-Effect Transistor (FinFET), Complementary Metal Oxide Semiconductor (CMOS), and Silicon on Insulator (SOI) with specific qualities such as Micro-Electro-Mechanical Systems (MEMS), Power Management, CMOS Image Sensor (CIS), Non-Volatile Memory (NVM), and Radio Frequency (RF) are some of the future requirements in the field of IoTGC.

4.3.3 Based on Processor

The CoreLH is a competent processor that is specially designed for various applications of IoTGC. The Core L and Core H of this processor are basically designed for various computation tasks that require less and high energy, respectively [18]. To minimize the utilization of the resources efficient scheduling methods are used. These scheduling methods allocate various tasks to each core of the processor. The manufactures of mobile phones face difficulty to add advanced features to smartphones due to the use of deep learning (DL) processor [48]. For example, in iPhone X a more precise face locking system can be implemented by integrating a neural engine with an A11 Bionic chip. Similarly, the advancement in neuromorphic processors can approximate the mechanism of human neurons [49]. The processor widely uses Spiking Neural Network (SNN) to make the system efficient in both the spatial and temporal domain. These processors consume less energy as compare to the deep learning (DL) processors and also resolve the problems associated with the online machine learning.

4.3.4 Based on Sensor

In IoTGC applications the use of sensors plays a vital role to detect the status of patient health and manage the security of smart homes. Sensors are also used as an integral part of various units of the industries. Proper utilization of sensors in various manufacturing units, sales units, and other sub-divisions significantly controls the production process and other important activities [11]. In the production department of the industries, these sensors are used to measure the pressure, temperature, and humidity. Different sensors like pressure sensors, motion sensors, thermometers, etc. are used to produce the electrical signals. These signals are further consumed by the computer for various important activities. Various classes [5, 20, 22, 45] of these sensors are given below.

- *Active or passive:* To boost up the work over the network, active sensors are widely used. These sensors require an external power source to carry out various important activities. The example of active sensors can be GPS and radar. Whereas the passive sensors do not require any additional external power sources, hence, they generate their own electrical signals. Examples of these types of sensors are electric field sensors, thermal sensors, and metal detecting sensors.
- *Based on property:* Different types of sensors can be categorized based on their usefulness and properties. Sensors are selected based on the immediate requirements in the field of automated systems, biomechanical systems, and other related areas.
- *Digital or Analog:* The sensors that can be used to generate and transfer analog and digital information. These sensors are used to transmit the information to other devices on the network.

To get the benefits of IoTGC applications, various types of sensors are integrated. These sensors are integrated based on their sensitivity, applicability, and reliability. By adopting the sensor-on-chips the power consumption of the sensors can be minimized. Utilization of these chips can reduce the traffic overhead and energy consumption that is required in communication. Similarly, other powerful sensors can be based on the requirements over the IoT network. Development of energy-efficient, cost-effective, and echo-friendly sensors is still a challenging task in the field of IoTGC.

4.4 Applications of IoT Green Computing Based on Awareness

IoT green computing has wide applications in various areas such as smart transportation systems, smart homes, etc. Still people do not aware of where IoT can function effectively. Due to unawareness among the users, they are not getting the benefits of IoT in their routine work. The precise feedback of various IoT-based applications is essential from the user to improve the existing IoT-based technologies [18].

The government must take the initiatives to begin a few awareness programs so that audiences, manufacturers, and service providers can include wide applications of IoTGC in their workplace. The effectiveness of these awareness programs depends upon the country, its cultures, and type of audience. The usefulness of the smart metering system must be communicated among the audience, house owners, and industries so that power can be utilized efficiently. Various users of IoTGC must be aware of the usefulness of energy control mechanism so that more efficient technologies can be developed by the industries. Various prominent areas of IoTGC are needed to be focused on the awareness programs which are shown in Fig. 10.

4.4.1 Usefulness of Ambient Notification

Fixed proliferous displays are widely used to aware the end-users about the effective use of green IoT or IoTGC. According to the market survey, it has been proven that the use of these devices for awareness is more effective than the smartwatch and smartphone. To fit the images of various awareness programs in these small screen display devices various image retargeting techniques are being developed [50–52]. Various devices are placed at different locations in the societies for the advertisement of IoTGC and to promote its various useful applications.

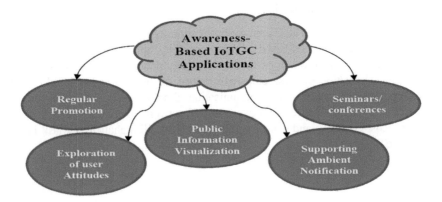

Fig. 10 Various IoTGC applications based on awareness

4.4.2 Proper Visualization of Information in Public Sectors

After adopting the techniques of information visualization the facts can be represented in a meaningful way. Representation of the facts in such a manner can increase the understanding of end-users about the IoTGC. The most common examples of information visualization can be dashboards and scatter plots. Information visualization provides useful information to the experts so that their skills can be utilized for a specific purpose. The concept of ambient visualizations can be adopted to influence various users who are interested to use IoT-enabled devices in their daily life. Various methods can be adopted to aware the public through information visualization. Representation of useful information through the decorated objects, visualization of information through air balloons, and mounting of ambient displays at various locations can significantly increase the awareness of IoTGC among the end-users [44, 45].

4.5 IoTGC Applications Based on Recycling

To attain successful environmental-friendly IoT green computing the raw materials used in the manufacturing of various products must be recycled. For example, non-biodegradable materials such as plastic, copper, or another similar kind of materials highly influence the greenhouse effect. These materials must be avoided in the development of cell phones and other electronic gadgets [18]. Similarly, instead of utilizing other power generating resources, solar energy in industries and homes can also minimize various environmental issues and challenges. Figure 11 shows various recycling-based IoTGC applications to minimize carbon footprints for better IoT green computing.

For energy-efficient IoTGC applications, the size of the network can be minimized by implementing efficient routing algorithms. Elimination of redundant data over the network can also save the energy that is required to access the nodes. The

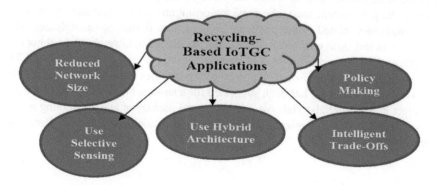

Fig. 11 Various steps used in recycle-based IoTGC

hybrid network architecture with various active and passive sensors should be used to reduce power consumption and to accomplish various IoTGC tasks. In this regard, concrete decisions, better planning, and policies can also reduce the power consumption in the manufacturing of various electronic products and construction of smart buildings [9]. Finally, various essential factors such as the cost, utilization of power, and other resources allow the smooth functioning of IoT green computing.

5 A Case Study on the Influence of Smartphones on the Environment

To implement the concept of green computing the products that are environmentally friendly must be manufactured. Green computing plays a major role to investigate the various influence of smart IoT-based devices such as mobile devices, computers, and other smart digital gadgets in the environment. With the exponential usage of these digital devices in the public domain, the problem of pollution in the environment can be faced. The heavy usage of smartphones can be the cause of the emission of CO_2 in the environment that can influence the present and future of mankind. This section presents a case study on smartphones. The main reason behind the selection of this highly used IoT component is its wide usage in the public domain. Through this case study, the impact of the smartphone on the environment and its recycling are discussed.

5.1 Minimization of the Drastic Change in the Environment

To maintain a sustainable environment manufactures are trying to incorporate various latest trends in the development of smartphones. To achieve green computing the smartphone must be assembled through energy-efficient hardware components. The government must take some major initiatives to investigate and develop new guidelines related to the hardware and software design of smartphones. The energy-efficient design of smartphones and the use of lightweight algorithms can minimize the energy consumption of smartphones. To minimize the release of CO_2 and its impact on the environment power saving batteries must be manufactured. In the available literature it has been suggested for a sustainable environment we need to select green materials. The cost incurred in the packaging and assembling of the smartphones can be minimized through the proper guidelines of the companies to their manufacturing units.

5.1.1 Hazardous Materials of Smartphone

Nowadays smartphones may contain various types of toxic materials. These materials can be nickel, mercury, lithium, etc. that are harmful to the environment. This section explains how these toxic materials are harmful to mankind. From the study, it has been revealed that the old smartphones must be recycled and should not be thrown in the garbage. The presence of Nickel in the batteries of mobile phones causes cancer. Similarly, in the batteries of older mobile phones, the presence of mercury found another type of dangerous substances. The presence of Magnesium may be responsible for the neurochemical changes. Manganese is widely used in the production of circuit boards of smartphones. The inconstancy property of lithium may cause various environmental problems. Most of the products may contain Zinc and Cadmium. The use of Cadmium at a high level may cause kidney failure. Arsenic that is a well-known poison is heavily used in the production of LEDs. Antimony compound that is poisonous and similar to arsenic can be widely used as a flame retardant. Beryllium material can be used in cooling may cause a prolonged allergic disease in the human body. Furthermore, due to the use of Copper, and other precious materials such as Iridium, Silver, and Gold companies are recycling smartphones for the production of new devices. Nowadays various mobile companies like Samsung, Motorola, Vivo, Oppo, etc. are promoting recycling programs through which people can exchange their old smartphones and can get some discount on a new device. From the available literature, it has been found that the PCBs of smartphones and heavy use of plastic material in the phone chargers may be harmful to the environment. The emission of CO_2 due to the burning of plastic material of smartphones may hamper the sustainability of the environment. It has been found that the printed circuit boards of smartphones are manufactured by 12% of polymers, 62% of metals, and 23% of ceramics [53].

5.1.2 Recycling

It is a process to apply new technologies to faulty materials to convert them into useful materials [54]. From the study [55] it has been found that in the market of China recycling of smartphones is very low. Most of the smartphones are sold after refurbishing. This kind of reuse of cellphones does not affect the sustainability of the environment. Most of the smartphone manufacturing companies are developing new guidelines for the refurbishment of the cellphones to develop a strong market. The results obtained from the study suggested that the first cell phone recycling act was developed in the state of California. Various policies under this act are mainly focused on the significance and positive impact of recycling on the environment. It has been noticed that approximately 17% of smartphones are available in the market for recycling and refurbishment that is a very less percentage. Government and other private agencies must initiate some awareness programs that can promote recycling and encourage the customers to understand the urgent need of smartphones waste management. China is promoting green card recycling activity and the green box environmental program [56]. Various activities in green IT can manage E-waste that can provide better solutions for the development of a sustainable environment.

5.1.3 Selection of Right Design of Repair and Disassembly

Most of the components of the smartphones are completely molded into the plastic case. These types of molding prevent the replacement of original components from the duplicate components available in the market. Designing and manufacturing of such smartphones may hamper the process of recycling. The smartphones can be broken for their refurbishment using electronic components that are inbuilt on their PCB. Such types of smartphones are not cost-effective as the extraction of reusable components is not possible by the customer.

5.1.4 Development of Smartphones using Green Materials

For the manufacturing of smartphones polylactic acid plastic (PLA) can be used. These materials are made up of corn starch or glucose that is renewable and biodegradable. Recycling of plastic materials and other green materials like bamboos can be used for the manufacturing of smartphones.

5.1.5 Energy-Saving Smartphone Batteries

The manufacturing companies must promote the wide use of natural and organic radical battery (ORB) for the production of smartphones. These batteries do not use heavy metals and can be charged within 30–40 seconds. The energy-saving capability of such types of batteries can reduce power consumption over the IoT network.

5.1.6 Reduce Packaging and Accessories Requirements

Mobile phone manufacturing companies must jointly take some initiatives to reduce the packaging cost of smartphones. In the packaging of the smartphones, various plastic materials and other useful substances are used. The manufacturing companies can manufacture common phone charger that can be used for all the brands of the smartphone. HTC, Nokia, and Sony now promote some units with simply USB lead alternatively of needless chargers.

5.2 Assessment of Smartphones using Life Cycle Assessment (LCA)

An environmental LCA (life cycle assessment) is a technique to assess the potential environmental aspects associated with the product and other services. This assessment can be applied for the analysis of the life cycle of smartphones. Through the analysis, the process of traditional manufacturing can be improved to minimize the hazardous impact of smartphones on the environment. Every 3 years LCA is used to analyze the impact of materials used in the smartphone on the environment. Figure 12 shows the life cycle of the LCA. Through the analysis of the results obtained from the LCA, it can be predicted that the requirements of power consumption and the exponential growth of the CO_2 emission can be significantly reduced up to 22–59% and 20–72%. LCA can also be used to analyze the results of the recycling process and material selection of the device in the environment.

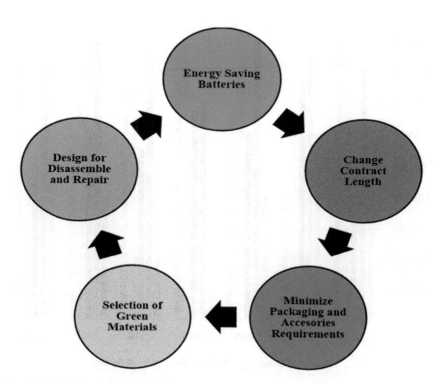

Fig. 12 Life Cycle assessment of Smartphones

5.3 Selling and Emission Rate of the Smartphone

With the exponential growth of the modern IT requirements, the manufacturing companies are expanding the structure of their manufacturing units. Due to this expansion smartphone companies are facing various challenges and issues related to green IT [57]. Figure 13 shows the emission of CO_2 in the environment due to heavy usage of toxic materials in smartphones. Table 2 shows various factors based on which LCA can be carried out that represent the impact of smartphones on the environment. After taking some important factors Table 3 shows the sales of some popular smartphones in the market and their impact on the environment in terms of CO_2 emission. The year-wise selling growth rate of the popular mobile phones is shown in Table 3. From Table 3, it can be observed that the selling growth rate of mobile is increasing that is responsible for the growth of the CO_2 in the environment. To forecast the accurate results the data of selling growth of the smartphones are collected from the year 2014 to 2020.

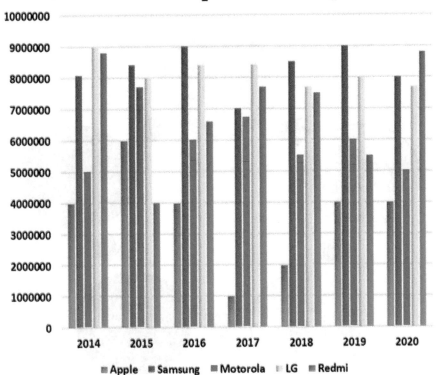

Fig. 13 Emission of CO_2 from Smartphones

Table 2 Computation of CO_2 percentage using LCA

S. No.	Technique	Influenced Medium	Important Factors	% of CO_2 Emission (%)	Ref. No.
1	To measure the performance of the Smartphones Life Cycle Assessment (LCA) is used.	AIR	Influence of smartphone chargers in life cycle	3	
2			Reduction in energy consumption	34	[58]
3			CO_2 emission in a complete life cycle of mobile phones	65–77	[59]
4			Percentage of CO_2 emission in 4 years through Sony XperiaTM smartphone	44	[60]
5			Due to use of Copper substances on the PCB of the smartphones.	33.6	[53]
6			Due to use of heat resistance components in the smartphones	49	[61]
7			Reduction of CO_2 emission before 2020.	19.5	[62]
8			Reduction of CO_2 emission in 2020	38.6	[63]

6 Discussion on Case Study

From the case study, it has been proved that the concept of IoT green computing comprised various techniques through which in a smart world a sustainable environment can be created. The exponential growth of IoT devices has changed the overall economy of the world. Due to exponential growth in the usages of smartphones researchers have been predicted that in 2025 out of 35 billion IoT devices can be connected to the Internet. The study carried out by Cisco Internet Business Group has been predicted that almost 8–9 IoT-enabled devices of a person are being connected to the Internet in 2020. From the literature, it has been predicted that the reduction of power consumption and emission of carbon footprints can be achieved through the development of new technologies. From the report obtained from the IEEE green ICT Initiatives, it has been predicted that ICT industries are responsible for the 4% of total CO_2 emissions. Figure 14 shows the year-wise exponential growth of IoT devices over the IoT network and their impact on the environment in terms of CO_2 emission.

Many manufacturing companies are trying to adopt new technologies to provide concrete solutions and strategies related to the problems. Through the case study, we

Table 3 Selling growth rate of smartphones

Smart Phone	2014 Units	2015 Units	2016 Units	2017 Units	2018 Units	2019 Units	2020 Units
Apple	1,914,258	2,258,506	2,602,756	2,947,002	329,125	3,635,498	3,979,746
Samsung	3,075,969	3,202,197	6,096,797	8,991,397	11,885,997	14,780,597	17,675,197
Motorola	185,660	204,460	223,314	242,168	261,022	279,876	298,730
LG	165,660	194,460	203,300	352,168	301,022	259,806	328,750
Redmi	175,656	177,560	182,375	212,100	12,900,240	16,822,600	11,565,730

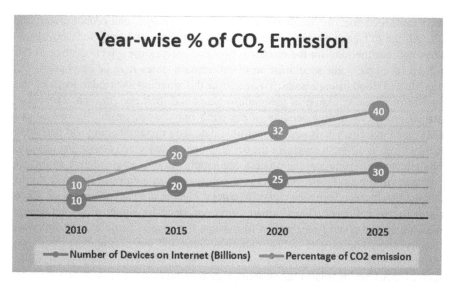

Fig. 14 Year-wise exponential growth of IoT devices and their impact on environment

provided a few observations and suggestions to the readers and researchers to find out new solutions for the growth of a sustainable environment.

Finally, through the above case study that has been carried out in Sect. 5 the following major points must keep in the mind in the production of smartphones.

- New strategies and solutions are required to find out for the development of the common hardware and software architectures in the production of smartphones.
- New recyclable materials are needed to be explored for the manufacturing of smartphones.
- Few policies and awareness programs must be initiated in the public sector through which IoT solutions can be deployed efficiently.

From the case study, we have investigated various possible solutions that can be deployed into real life to achieve environment friendly and energy efficiency green IoT. The study carried out in the case study reveals various alarming situations that are needed to take care of. The ignorance towards possible challenges can hamper the growth of a sustainable environment.

7 Conclusion

Nowadays industries are adopting the environment in which the demand of IoT devices is increasing exponentially. The purpose of implementing green IoT is to minimize the challenges and issues that can hamper the environment sustainability. For the establishment of a greener and smart world, IoT devices should be easily acceptable, eco-friendly, easily disposable, energy-efficient, and economically viable. All the IoT-enabled devices needed ample amounts of energy for their operations and put the load on the electrical grid and greenhouse gases. Change in the pattern of climate and accumulation of greenhouse gases may be responsible for drought and flood in some areas. To overcome this situation we need to focus on the growth of global emission of greenhouse gases and their preservation. The exponential growth in the temperature of the earth may be responsible for the severe environmental glitches. Based on real-time data the crucial planning and decisions can be taken at different levels for the development of IoT green computing devices. The effective decision-making process can supply various inputs to give an impact on the environment. Proper promotion of policies and awareness of the green world among the human being can play a crucial role in the development of green campuses or residence. These smart campuses must be equipped an automated monitoring and control mechanism to manage the several facilities economically. In conventional health care and agriculture systems, IoT green computing can play a major role. IoT green computing-based health care systems have some applications such as smart health monitoring systems, handling the massive amount of data using data mining and artificial intelligence techniques. Similarly, a smart agriculture system enables farmers to increase crop yields, and also aware farmers about the efficient technology of irrigation. Through IoTGC based smart vehicle systems, time and cost can be significantly saved. In these automated systems various IoT-based sensors are used to detect the possibility of an accident in advance and also provide e-notification about traffic and weather conditions via remote servers. The IoTGC can be used to design the map of intelligent houses that can provide comfort to the end-users. Smart homes provide various facilities to the users so that they can be capable to manage their daily scheduled and plans. Due to significant improvements in the World Wide Web the human being and hardware devices are now linked together at every place and time. The smart security system based on IoT can provide a way to disseminate sensitive information among various private and government agencies. In the IoTGC system, efficient software and hardware design play a crucial role to make the system eco-friendly and efficient in terms of power consumption. The government must take the initiative to begin a few awareness programs so that audiences, manufacturers, and service providers can include wide applications of IoTGC in their workplace. The effectiveness of these awareness programs depends upon the country, its cultures, and type of audience. To attain successful environmental-friendly IoT green computing the raw materials used in the manufacturing of various products must be recycled. For example, non-biodegradable materials such as plastic, copper, or other similar kinds of materials highly influence the greenhouse effect. Researchers are continuously providing their contribution to various applications of IoTGC based on some important factors that are essential for the establishment of the smart world. Efficient technologies, protocols, and algorithms can play a significant role in the field of smart health care monitoring systems, smart vehicle monitoring systems, smart manufacturing systems, etc. The human being can make their life luxurious by adopting various sensors, actuators, and other useful IoT-enabled devices. Based on various highlighted issues and challenges related to different categories such as policy, hardware, software, awareness, and recycling the potential researchers can contribute their significant endeavors to improve various aspects of IoT and green computing (GC). To justify the study of this chapter a case study has been carried out on the development of smartphones. The case study suggests major points that must be kept in the mind during the development of smartphones. From the case study, we have investigated various possible solutions to achieve environment friendly and energy efficiency green IoT. The case study reveals various factors that are needed to take care of. The ignorance of these suggested challenges can hamper the growth of a sustainable environment.

References

1. Atzori, L., Antonio, I., & Giacomo, M. (2010). The internet of things: A survey. *Computer Networks, 54*(15), 2787–2805.
2. Miorandi, D., Sabrina, S., De Pellegrini, F., & Chlamtac, I. (2012). Internet of things: Vision, applications and research challenges. *Ad Hoc Networks, 10*(7), 1497–1516.
3. Al-Fuqaha, A., Guizani, M., Mohammadi, M., Aledhari, M., & Ayyash, M. (2015). Internet of things: A survey on enabling technologies, protocols, and applications. *IEEE Communications Surveys and Tutorials, 17*(4), 2347–2376.
4. Botta, A., De Donato, W., Persico, V., & Pescapé, A. (2016). Integration of cloud computing and internet of things: A survey. *Future Generation Computer Systems, 56*, 684–700.
5. Chase, J. (2013). The evolution of the internet of things. *Texas Instruments, 1*, 1–7.
6. Rani, S., Talwar, R., Malhotra, J., Ahmed, S. H., Sarkar, M., & Song, H. (2015). A novel scheme for an energy efficient Internet of Things based on wireless sensor networks. *Sensors, 15*(11), 28603–28626.
7. Sathyamoorthy, P., & Ngai, E.. (2015). Energy efficiency as a orchestration service for the Internet of Things. In *Proc. 11th Swedish Nat. Comput. Netw. Workshop* (pp. 41–44).
8. Visalakshi, P., Paul, S., & Mandal, M. (2013, May). Green computing. In *Proceedings of the National Conference on Architecture, Software systems and Green computing (NCASG), Paiyanoor (India)* (pp. 63–69).
9. Nandyala, C. S., & Kim, H. K. (2016). Green IoT agriculture and healthcare application (GAHA). *International Journal of Smart Home, 10*(4), 289–300.
10. Kallam, S., Madda, R. B., Chen, C. Y., Patan, R., & Cheelu, D. (2017). Low energy aware communication process in IoT using the green computing approach. *IET Networks, 7*(4), 258–264.
11. Mekala, M. S., & Viswanathan, P. (2020). A survey: energy-efficient sensor and VM selection approaches in green computing for X-IoT applications. *International Journal of Computers and Applications, 42*(3), 290–305.
12. Solanki, A., & Nayyar, A. (2019). Green internet of things (G-IoT): ICT technologies, principles, applications, projects, and challenges. In *Handbook of research on big data and the IoT* (pp. 379–405). IGI Global.
13. Nayyar, A., Puri, V., & Le, D. N. (2017). Internet of nano things (IoNT): Next evolutionary step in nanotechnology. *Nanoscience and Nanotechnology, 7*(1), 4–8.
14. Xiaojun, C., Xianpeng, L., & Peng, X. (2015). IOT-based air pollution monitoring and forecasting system. In *IEEE international conference on computer and computational sciences (ICCCS)* (pp. 257–260).
15. Da Xu, L., He, W., & Li, S. (2014). Internet of things in industries: A survey. *IEEE Transactions on Industrial Informatics, 10*(4), 2233–2243.

16. Kamilaris, A., & Pitsillides, A. (2016). Mobile phone computing and the internet of things: A survey. *IEEE Internet of Things Journal, 3*(6), 885–898.
17. Curry, E., Guyon, B., Sheridan, C., & Donnellan, B. (2012). Developing a sustainable it capability: Lessons from Intel's journey. *MIS Quarterly Executive, 11*(2), 61–74.
18. Arshad, R., Zahoor, S., Shah, M. A., Wahid, A., & Yu, H. (2017). Green IoT: An investigation on energy saving practices for 2020 and beyond. *IEEE Access, 5*, 15667–15681.
19. Mineraud, J., Mazhelis, O., Su, X., & Tarkoma, S. (2016). A gap analysis of Internet-of-Things platforms. *Computer Communications, 89*, 5–16.
20. Wang, H. I. (2014). Constructing the green campus within the internet of things architecture. *International Journal of Distributed Sensor Networks, 10*(3), 804627.
21. Wang, K., Wang, Y., Sun, Y., Guo, S., & Wu, J. (2016). Green industrial Internet of Things architecture: An energy-efficient perspective. *IEEE Communications Magazine, 54*(12), 48–54.
22. Moreno, M., Úbeda, B., Skarmeta, A. F., & Zamora, M. A. (2014). How can we tackle energy efficiency in IoT based smart buildings? *Sensors, 14*(6), 9582–9614.
23. Keertikumar, M., Shubham, M., & Banakar, R.M. (2015). Evolution of IoT in smart vehicles: An overview. In *2015 IEEE International Conference on Green Computing and Internet of Things (ICGCIoT)* (pp. 804–809).
24. Kamal, M. A. S., Imura, J. I., Hayakawa, T., Ohata, A., & Aihara, K. (2014). Smart driving of a vehicle using model predictive control for improving traffic flow. *IEEE Transactions on Intelligent Transportation Systems, 15*(2), 878–888.
25. Zamora-Izquierdo, M. A., Santa, J., & Gómez-Skarmeta, A. F. (2010). An integral and networked home automation solution for indoor ambient intelligence. *IEEE Pervasive Computing, 9*(4), 66–77.
26. Gapchup, A., Wani, A., Wadghule, A., & Jadhav, S. (2017). Emerging trends of green IoT for smart world. *International Journal of Innovative Research in Computer and Communication Engineering, 5*(2), 2139–2148.
27. Doknić, V. (2014). Internet of things greenhouse monitoring and automation system. Internet of things: smart devices, processes, services.
28. Prasad, R., Ohmori, S., & Simunic, D. (Eds.). (2010). *Towards green ICT* (p. 9). River Publishers.
29. Peoples, C., Parr, G., McClean, S., Scotney, B., & Morrow, P. (2013). Performance evaluation of green data centre management supporting sustainable growth of the internet of things. *Simulation Modelling Practice and Theory, 34*, 221–242.
30. Perera, C., Talagala, D. S., Liu, C. H., & Estrella, J. C. (2015). Energy-efficient location and activity-aware on-demand mobile distributed sensing platform for sensing as a service in IoT clouds. *IEEE Transactions on Computational Social Systems, 2*(4), 171–181.
31. Murugesan, S. (2008). Harnessing green IT: Principles and practices. *IT Professional, 10*(1), 24–33.
32. Rohokale, V. M., Prasad, N. R., & Prasad, R. (2011). A cooperative Internet of Things (IoT) for rural healthcare monitoring and control. In *2011 2nd IEEE international conference on wireless communication, vehicular technology, information theory and aerospace & electronics systems technology (Wireless VITAE)* (pp. 1–6).
33. Kim, J. (2015). Energy-efficient dynamic packet downloading for medical IoT platforms. *IEEE Transactions on Industrial Informatics, 11*(6), 1653–1659.
34. Palo, H. K., Chandra, M., & Mohanty, M. N. (2018). Recognition of human speech emotion using variants of Mel-Frequency cepstral coefficients. In *Advances in systems, control and automation* (pp. 491–498). Springer.
35. Palo, H. K., Kumar, P., & Mohanty, M. N. (2017). Emotional speech recognition using optimized features. *International Journal of Research Electronics Computer Engineering, 5*(4), 4–9.
36. Galinina, O., Mikhaylov, K., Andreev, S., Turlikov, A., & Koucheryavy, Y. (2015). Smart home gateway system over Bluetooth low energy with wireless energy transfer capability. *EURASIP Journal on Wireless Communications and Networking, 1*, 178.

37. Gou, Q., Yan, L., Liu, Y., & Li, Y. (2013). Construction and strategies in IoT security system. In *2013 IEEE international conference on green computing and communications and IEEE internet of things and IEEE cyber, physical and social computing* (pp. 1129–1132).
38. Shi, W., Cao, J., Zhang, Q., Li, Y., & Xu, L. (2016). Edge computing: Vision and challenges. *IEEE Internet of Things Journal, 3*(5), 637–646.
39. Sun, X., & Ansari, N. (2016). EdgeIoT: Mobile edge computing for the Internet of Things. *IEEE Communications Magazine, 54*(12), 22–29.
40. Yi, S., Hao, Z., Qin, Z., Li, Q. (2015). Fog computing: Platform and applications. In *2015 Third IEEE Workshop on Hot Topics in Web Systems and Technologies (HotWeb)* (pp. 73–78).
41. Atlam, H. F., Walters, R. J., & Wills, G. B. (2018). Fog computing and the internet of things: A review. *Big Data and Cognitive Computing, 2*(2), 10.
42. Al-Azez, Z. T., Lawey, A. Q., El-Gorashi, T. E., & Elmirghani, J. M. (2015). Virtualization framework for energy efficient IoT networks. In *2015 IEEE 4th International Conference on Cloud Networking (CloudNet)* (pp. 74–77).
43. Colella, R., Catarinucci, L., & Tarricone, L. (2016). Improved RFID tag characterization system: Use case in the IoT arena. In *2016 IEEE International Conference on RFID Technology and Applications (RFID-TA)* (pp. 172–176).
44. Mirlacher, T., Buchner, R., Förster, F., Weiss, A., & Tscheligi, M. (2009). Ambient rabbits likeability of embodied ambient displays. In *European conference on ambient intelligence* (pp. 164–173). Springer.
45. Mikusz, M., Houben, S., Davies, N., Moessner, K., & Langheinrich, M. (2018). *Raising awareness of IoT sensor deployments*.
46. Singh, A. K., & Patro, B. D. K. (2020). Elliptic curve signcryption based security protocol for RFID. *KSII Transactions on Internet & Information Systems, 14*(1). https://doi.org/10.3837/tiss.2020.01.019
47. Singh, A. K., Solanki, A., Nayyar, A., & Qureshi, B. (2020). Elliptic curve signcryption-based mutual authentication protocol for smart cards. *Applied Sciences, 10*(22), 8291.
48. Mohammadi, M., Al-Fuqaha, A., Sorour, S., & Guizani, M. (2018). Deep learning for IoT big data and streaming analytics: A survey. *IEEE Communications Surveys & Tutorials, 20*(4), 2923–2960.
49. Venkataramani, S., Roy, K., & Raghunathan, A. (2016). Efficient embedded learning for IoT devices. In *2016 IEEE 21st Asia and South Pacific Design Automation Conference (ASP-DAC)* (pp. 308–311).
50. Garg, A., & Negi, A. (2020). Structure preservation in content-aware image retargeting using multi-operator. *IET Image Processing, 14*(13), 2965–2975.
51. Garg, A., Negi, A., & Jindal, P. (2020). Structure preservation of image using an efficient content-aware image retargeting technique. Signal. *Image and Video Processing*, 1–9.
52. Garg, A., & Negi, A. (2020). A survey on content aware image resizing methods. *KSII Transactions on Internet and Information Systems (TIIS), 14*(7), 2997–3017.
53. Yamane, L. H., de Moraes, V. T., Espinosa, D. C. R., & Tenório, J. A. S. (2011). Recycling of WEEE: Characterization of spent printed circuit boards from mobile phones and computers. *Waste Management, 31*(12), 2553–2558.
54. Rathore, P., Kota, S., & Chakrabarti, A. (2011). Sustainability through remanufacturing in India: A case study on mobile handsets. *Journal of Cleaner Production, 19*(15), 1709–1722.
55. Yu, J., Williams, E., & Ju, M. (2010). Analysis of material and energy consumption of mobile phones in China. *Energy Policy, 38*(8), 4135–4141.
56. Zhou, L., & Xu, Z. (2012). Response to waste electrical and electronic equipments in China: Legislation, recycling system, and advanced integrated process. *Environmental Science & Technology, 46*(9), 4713–4724.
57. Khan, S. U., Wang, L., Yang, L. T., & Xia, F. (2013). Green computing and communications. *The Journal of Supercomputing, 63*(3), 637–638.
58. Sun, C. (2012). Application of RFID technology for logistics on internet of things. *AASRI Procedia, 1*, 106–111.

59. Moberg, Å., Borggren, C., Ambell, C., Finnveden, G., Guldbrandsson, F., Bondesson, A., Malmodin, J., & Bergmark, P. (2014). Simplifying a life cycle assessment of a mobile phone. *The International Journal of Life Cycle Assessment, 19*(5), 979–993.
60. Ercan, E. M. (2013). *Global warming potential of a smartphone: Using life cycle assessment methodology*.
61. Chowdhury, C. R., Chatterjee, A., Sardar, A., Agarwal, S., & Nath, A. (2013). A comprehensive study on cloud green computing: To reduce carbon footprints using clouds. *International Journal of Advanced Computer Research, 3*(8), 78–85.
62. Tushi, B. T. (2015). An archival analysis of green information technology: The current state and future directions. *Doctoral dissertation, Queensland University of Technology*.
63. Samuri, N. (2014). Making green IT "alive" at TVET Institution of Malaysia. In *Proc. 2nd Int. Conf. Green Comput. Technol. Innov.(ICGTI)* (pp. 12–18).

Vehicular Intelligence System: Time-Based Vehicle Next Location Prediction in Software-Defined Internet of Vehicles (SDN-IOV) for the Smart Cities

Preeti Rani, Naziya Hussain, Rais Abdul Hamid Khan, Yogesh Sharma, and Piyush Kumar Shukla

1 Introduction

Across the preceding two decades, with the improved modern mechanization, we have experienced vast development in smart devices, we can try to log on to network facilities and applications. Further, the basis of a system that connects such devices has stayed constant since its invention. The fact is that the need for people, gadgets, and resources to use a network is increasingly growing over time. Network function virtualization (NFV) and software-defined networking (SDN) are paired tactics while presenting a distinctive approach to layout and administer the system. SDN mechanization offers a program for assessing and executing a new groundbreaking concept while discovering its programmability and integrated management system.

It divides the data plane from the control plane to provide a unified view of the scattered system [1]. IOV is another mechanization that expands swiftly. The state organizations determine their exercises. This useful communication structure

---

P. Rani (✉)
SRMIST, Delhi-NCR Campus, Modinagar, Ghaziabad, India

N. Hussain
School of Computers, IPS Academy, Indore, India

R. A. H. Khan
G.H. Raisoni University, Amravati, India
e-mail: rais.khan@ghru.edu.in

Y. Sharma
Vishwakarma Institute of Information Technology, Pune, India
e-mail: yogesh.sharma@viit.ac.in

P. K. Shukla
UIT, RGPV, Bhopal, India

© The Author(s), under exclusive license to Springer Nature Switzerland AG 2021
F. Al-Turjman et al. (eds.), *Intelligence of Things: AI-IoT Based Critical-Applications and Innovations*,
https://doi.org/10.1007/978-3-030-82800-4_2

promotes the enhancement of the intelligent transportation system (ITS). The characteristic of the IOV consists of various expertise connected with fastest internet, inevitable agility, and changeable network density [2–5], that is not obtainable in mobile ad-hoc network (MANET), In which inadequate battery capacity, casual gesture, and calculation. IOV is unalike from the VANETs by maintaining a unified organization methodology, making it more appropriate for ITS security submission. IOV is developing as a hopeful ahead, the locked and patented way of supervising network tools nowadays. We trust that SDN's advantages can link the break among the path safety systems and IOV. A prolonged form of SDN into IOV is shown below in Fig. 1.

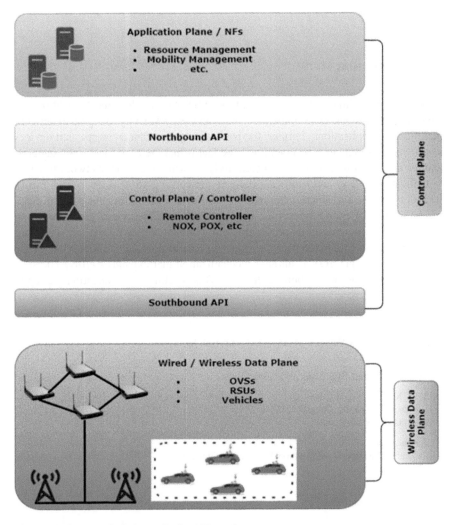

Fig. 1 The Layout of a Software-Defined Network

Over the past decade, the ITS has progressed dramatically, consisting of numerous telecommunications in an informal network called VANET, such as V2V (Vehicle to Vehicle), V2R (Vehicle to Roadside), and V2I (Vehicle to Infrastructure) [6, 7]. Nonetheless, the mechanical limits to support many cars and the very little commercial model for VANET have inadequate placement, creating the IOV. With expanded abilities, IOV is an addition to VANET. It is a commercial-oriented building that includes internet and mixed wireless access (WP) networks and mega target areas (protection, optimization of traffic and productivity, documentary, etc.). For example, IOV could generate a return of the latest automation to enable wider and all-in-one positioning; Mobile Edge, Cloud computing, Virtualization and Fog Computing, etc. Such cloud and fog computing, for example, is taken for the immediate conclusion in attractive responses in self-directed vehicles.

The new model SDN has interested scholars to influence its versatility and programmability in resolving IOV encounters, mainly Quality of Service (QoS) and scalability problems inherited from vehicular.

In this chapter, an Efficient Routing Algorithm for SDN-IOV is proposed by the name of ERS-SDN-IOV. The proposed ERS-SDN-IOV approaches are used the edge controllers (EC) to gain the assistances of centralized and decentralized information or data packet handling. The main contributions of the proposed ERS-SDN-IOV approaches are mentioned below.

1. The proposed ERS-SDN-IOV approaches used to estimate the shortest and more stable route among the vehicles.
2. It is used to predict the vehicle future location.
3. The location prediction model of vehicles is designed based Artificial Neural Network (ANN) on the dataset. The dataset is generated based on the vehicle activity such as moving speed of the vehicle, direction, junction, and road ID of the vehicle.
4. The future location prediction model of the moving vehicles is used to improve the efficiency of the routing system.

This chapter is arranged as follows: literature review of SDN, SC, and vehicular network communication is given in Sect. 2. Section 3 discusses the proposed methodology and Software-Defined Vehicular Network (SDVN) architecture and experimental results with simulation parameters are mentioned in Sect. 4. Section 5 demonstrates the conclusion and future scope.

2 Literature Review

VANETs have fascinated plenty of researchers in the past few years. Even though VANETs are organizing in certainty by providing numerous facilities, the existing style has been fronting several positioning and administration problems because of weak connectivity, less scalability, lesser suppleness, and lesser cleverness [8]. This Paper proposes a new VANET model called FSDN that syndicates SDN and fog computing as a forthcoming response to the development of the calculation and device model. The SDN-based design provides versatility, programmability, scalability, and world-wide knowledge [9–12]. Simultaneously, fog computing suggests amenities for location-awareness and delay-sensitive that can satisfy the strains of upcoming VANET circumstances [13–17].

Anwer and Guy [18] indicated that by increasing V2V, V2I, V2BS (Vehicle-to-Base Station) system, as well as SDN centralized control, thus maximizing the service of resources and reducing latency through the implementation of fog computing, the design could solve the critical challenges in the VANETs. Two use cases are also provided for non-safety (data streaming) and safety (Lane-change assistance) services to demonstrate the advantages of our proposed system.

Automated cars have converted into the latest scientific development in the transport sector. They anticipated transforming carriage by creating it securer, comfortable, and more resourceful. Nevertheless, there are several technical problems with the safety of automated cars and the disastrous possessions these safekeeping matters can have on the community. They had appropriate safety procedures in settlement for those cars stances a problem to their implementation for people. That is one of the main constraining aspects. The writer deliberates the self-directed expertise of automated cars and the safety issues with this technology. Discovering the existing communiqué skills of vehicular informal schmoozing and how non-artificial neural systems and software-based networking can help defend those cars from malicious attacks [19].

In the past few years, VANETs have been observed as empowering techniques to deliver a varied range of amenities, for instance, driving safety, boosted traffic and transportable efficacy, and ease for travelers and chauffeurs. The system (VANET) needs suppleness in its design that does the positioning of amenities/etiquettes more broadly.

Ydenberg et al. [20] show how SDN, an emerging system model, can be used to provide networks with flexibility and programmable functionality and provide VANETs with new services and options. Specifically, we take SDN's idea of wire substructures in the Data Center Room and suggest SDN-based VANET and its operating approach to adapt SDN to VANET system. In which a SDN-VANET and the facilities that can be provided are often intentionally advantaged [21, 22]. In contrast, SDN-based routing with old-design MANET/VANET protocols proves in imitation the feasibility of a SDN-VANET [23, 24]. It is also demonstrates in imitation fullback technique that should be delivered to relate the SDN concept to portable wireless set-ups, and displays one of the possible facilities that a Software-Defined VANET [25].

Ji et al. [26] proposed the SDN-based geographic routing (SDGR) manuscript procedure for VANET, according to prominence location, compactness of vehicles, and numerical plot. To dissociate network management from information shifting SDN is used. In SDGR, the chief manager collects data and info from automobiles and affords an international interpretation to calculate the best direction-finding tracks. Reproduction outcomes display that the projected scheme executes much greater than associated steering procedures in tenure of equal to the package distribution percentage and distribution interruption period.

SDN is a developing ability that differentiates the regulator plane from the advancing plane of knowledge in modifications and gathers all control planes into a chief operator. The central controller gathers network information from changes in the SDN-based routing plan and determines the best guiding tracks for modifications based on detailed web data and materials from around the world [25]. When adjustments do not need to interchange steering materials and data with each other, the operating cost is so lesser. The first is to propose an SDN-based map-reading outline for excellence communication dissemination in VANET. The latest procedure is established to discover the international most acceptable direction from the basis to the target in VANET with energetic system density, validate over the imitation outcomes, that projected agenda influential outdoes the relevant procedures about both distribution postponement period and direction-finding operating cost [27].

ONOS (Open Network Operating System), an investigational disseminated SDN manage program encouraged by the accomplishment, scalability, and accessibility needs of massive operative systems, define and assess two ONOS models. The initial edition applied essential elements: a supplied, although rationally integrated, worldwide-web view; scale-out; and error forbearance. The other edition concentrated on enhancing the presentation. Depend on familiarity with models as mentioned earlier, detect extra measures that will be needed for ONOS to favor utilize instances, such as leading system traffic engineering and timetabling, and turn into a functional publicly available, dispersed network OS type SDN [28].

By way of the forthcoming tendency of urban improvement, the smart city (SC) can offer accessible facilities for citizens [31]. The SC includes various areas, such as municipal public utilities, residential residents, transportation management, prescription for health control, grocery shopping, safety reassurance, etc. Nowadays, with the expansion of the IoT, several goods/ elements of SCs are connected with doing tasks in a co-operative fashion. The SH can offer more thorough facilities [32], for instance, power organization [33], IOV for smart cities [34], IOT enabled electric cars in smart cities [35]. Fog computing depends on the Internet of Vehicles (F-IOV) [36] (Fig. 2).

At present, many smart devices are connected with Cyberspace, and IoT automation is in authorizing numerous requests in the SC [37]. Besides, some important items are also incorporated in SCs for physical instruments (e.g., properties of items, data produced by devices, and personal qualities).

As shown in Fig. 1, the SC may include several conditions, with the Smart Grid (SG), ITS, Shopping Recommender System (SRS), Intelligent Medical Diagnosis (IMD), and so on. To explain more, the SG will increase the source of energy to decrease the overall consumption of electricity. The ITS can offer the safest journey path for passengers. The IMD, then again, can recommend acceptable health check policies permitting the sufferer's illness. Lastly, the SRS can propose suitable manufactured goods matching to consumer's requirements. It is observed that

Fig. 2 Smart city scenarios

these normal appliance situations in SC are centered on the intellectual communication and handling of a huge volume of data gathered from numerous gadgets/items in the SC. Furthermore, the sensible facilities that help customers make wise conclusions are progressively widespread because the QoE is progressively significant in SC applications [38]. Nevertheless, rational facilities' requirement needs an exceptionally solid information managing ability, which inspires the artificial intelligence (AI) inspired structure in the SC [39].

Li et al. [40] have established a software called Urban Mobile Sensing to observe sound, air, electromagnetic areas as a facility. This software is according to the MCS model and can accumulate SC data to increase the lifetime value for residents and allow civic leaders to draw conclusions [41], submitted an MCS software, and it influences people by applying the k-CLIQUE algorithm to improve people's involvement in MCS projects [42], explored the distributed sensor network usage for smart city applications and submitted SC detection techniques covering dedicated and non-dedicated sensors. The dedicated sensors are intended for some purposes, and similar smart devices are used to build non-dedicated sensors [43], proposed a portable detection method for the assessment of traffic compactness.

Various researches have been done work on vehicular network using SDN paradigm. The few research works listed here is designed based on the IOV integrated with SDN, digital map, and Fog concept, as shown in Table 1.

Table 1 Literature Comparisons

Author	Year	SDN	Digital map	Edge/Fog concept	Position prediction
[19]	2015	Yes	No	Yes	No
[26]	2016	Yes	Yes	No	No
[25]	2014	Yes	No	No	No
[1]	2020	Yes	Yes	Yes	No
[29]	2017	Yes	Yes	No	No
[27]	2015	Yes	No	Yes	No
[30]	2020	Yes	Yes	Yes	Yes

3 Proposed Methodology

This section discusses our proposed approach layout with their components in detail.

3.1 Proposed (ERS-SDN-IOV) Routing Protocol

A routing practice known as Hybrid Software Defined Networking Geographic Routing Algorithm suggests an SDN-based steering procedure with a partially dispersed command plane in this section (HSDN-GRA). On every lump of the vehicle system, the transmission control and steering regulations are applied, ensuring the constant entrance to the sources and the accuracy and originality of the data. SDN and network function virtualization (NFV) mechanisms suggest possible resolutions to reach supple and automatic grid administration, world-wide system optimization, and well-organized network source orchestration with inexpensiveness and are projected as a key enabler to forthcoming IOV. SDN is a pledging networking mechanism to make simpler network-layer information sending and enhance network-level source allotment. The data plane comprises several RSUs and motor vehicles and is deemed OVS.

Open Flow was the initial open-source management procedure for transmitting among the SDN controller and the system machines. Open Flow allows the enactment of an operator application platform to control system devices deprived of executing a variety of network procedures. Besides, on the network computer, Open Flow keeps the flow chart (forwarding devices). The movement map provides specifics of the way the data needs to be sent. By changing this motion map, the SDN controller can then use Open Flow to program an Open Flow-enabled switch's network resources. The Open Flow architecture allows two types of functions, called reactive and proactive, to plan the sending data and information and construct the system's route. The reactive method is the welshing technique of using Open Flow to execute SDN and presumes that there is no brainpower passing on the network resources from a control layer. In this method, the first data-traffic package received from other vehicle is forwarded to the SDN controller, after which this information is received by the SDN controller to programed movement across the entire system.

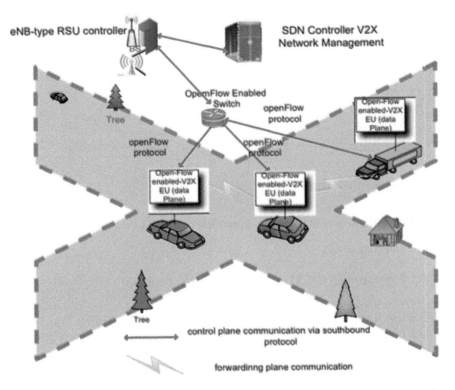

Fig. 3 The Software-Defined Vehicular Network (SDVN) architecture with the vehicle, roadside unit (RSU) on data plan

The SDN controller is pre-configured in hands-on mode with some default flow numbers, and traffic movement is immediately planned pre-emptively.

The elements of numerous wireless links in the SDN-VANET are shown in Fig. 3. To make the SDN-based vehicle network (SDVN) available, the imitation that is run on it directs to a specific number of SDN elements. The key sensible brainpower of the SDN-based vehicle structure is the SDN system. The SDN controller has a detailed and global definition of the vehicle network and applies the Open Flow procedure to handle the eNB-type RSU interface routing guidelines on the RSU. Indeed, the eNB-type RSU controller located on the side of the vehicle network decreases the option of producing new indeterminate forwarding packages in the flow charts of dispatching units. The administration of the SDN controller V2X system communicates routing instructions to vehicles by executing ITS goals built on the cloud or edge of the system to minimize handling decisions. The SDN controller is responsible for giving the entire presentation and provides gadgets (vehicles) with forwarding regulations that pick the safest routing route to their VANET locations. The Open Flow enabled by V2X-EU is the SDN vehicle and is responsible for controlling the components of the data plane. The Open Flow protocol is implemented by the vehicle data plane and installed in the On-Board

Diagnostic Device (OBU). In addition, data plane components are the VUEs that control note for routing guidance as of the eNB-type RSU controller to implement arranged steps that state the goals of ITS after being implemented in the SDVN request plane [35].

3.2 Prediction Neural Network

To predict different performance levels described in Fig. 4, artificial neural networks (ANNs) can be implemented. The benefit of ANNs involves automated learning only from the calculated data dependencies without any requirement to add more knowledge also, they can only learn through examples, and they can capture hidden and strongly nonlinear dependencies after their learning is done, even when there is significant noise in the training dataset.

To uncover secret dependencies and be able to use them to forecast the future, ANN is equipped to use historical data. In other words, with an explicitly given model, the neural network is not represented, and it has been described as a black-box learning approach. With time series, it is possible to predict several kinds of data. The time series indicates the time value, and this value can be influenced by other variables other than time. The time series reflects a discrete value background, which can be obtained by sampling from a continuous function [44, 45].

The Efficient Routing System (ERS) is the essence of the entire structure and is responsible for creating the enhanced end-wise route. This component connects with each other components. Together with that, it finds automotive information and data from the network topology table and recognizes existing automotive locations, speed, emphasis, and some other constraints. The Machine Learning Model (MLM) utilizes this routing system to forecast a motor vehicle's status using the Neural Network system. To have road data in progress, the ERS also connects with a Geographical Information System (GIS).

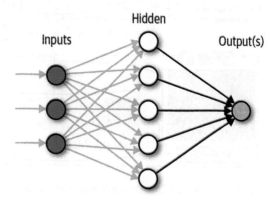

Fig. 4 The architecture of the Artificial Neural Network

Using a set of hop calculations, ERS then obtains the end-wise paths and makes a list in climbing order. ERS calls the MLM functionality after making a list by transmitting the paths with each motor vehicle's information on the route chosen.

The MLM unit assumes a situation and returns the predicted results to the ERS unit, where the system's forthcoming topology is predictable. The connectivity of the road and, accordingly, a long path is also tested and calculated by ERS. Suppose the life expectancy is greater than the intensity of the upper limit. In that case, the path is selected, or else the ERS again uses the same approach from the fastest route directory as the next chosen route. After calculating the shortest and most secure route, the ERS module integrates additional flow rules into the flow charts of the corresponding vehicles and RSUs only, which are included in the route chosen.

Every (Edge Controller) EC is accountable for covering up a particular region so as to obtain the information and data to decrease the web traffic flow operating cost. Therefore, in this case, with only 8-10 RSUs, every EC transmits. EC initially checks its target motor vehicle database every time there is a need for a new-found path, and a fast and steady path is expected for it. Since the target car is outside the EC's range, a request is sent to the SDN computer. SDN has the entire situs map for web analysis, so SDN can use an efficient method to measure the end-path. The EC launch also helps handle simultaneous data flow, minimizing the number of calls transmitted to the SDN controller from motor vehicles.

Machine Learning Module (MLM) is very important for anticipating the automobile location in our planned prototype. Figure 5 illustrates the suggested MLM interior design for our directing methodology. For location forecast, ANN is

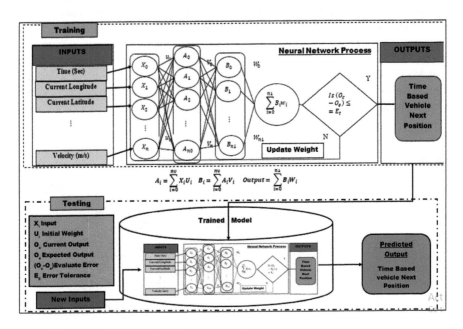

Fig. 5 The proposed ERS-SDN-IOV proposed system layout

discovered. ANN is a web of linked non-natural nerve cells that are prearranged in layers. There are three kinds of layer: input, output, and numerous hidden layers them. In each layer, various nerve cells are accessible for the duration of the preparation. Nerve cells in the concealed coats renew their weights randomly to boost weights wherever the mistake is reduced. When a version is guided, it could be utilized additional with additional items of input to forecast the automobile prospect situation on roads. Together with the purpose to deliver enhanced expectations, various versions for each side device are guided to their corresponding region only. This is because each guided version concentrates only on a specific area based on its traffic flow and its performance, outcomes in superior situation projection.

4 Experimental Results

4.1 Simulation Parameters

The simulation set-up of the network simulator NS-3 [46] for an intelligent vehicle system is described in this section. First, in order to assess performance, used SUMO [47] to produce vehicle tracks in the urban background and that scenario used for experiments. A city map with highways (map.osm) is exported from the Open Street Map (OSM) [47] to cover an area of (2000 x 2500) meters in order to perform the simulations. To randomly generate the traffic routes, we then used a python script using the randomTrips.py script given by the SUMO organization in the SUMO software that generates random traffic. It randomly assigns the environment to random velocity, direction, and courses that are nearer to more true. The next move is to link the SUMO with NS3 to mechanism the mobility and synchronize with the vehicular topology. Using an intelligent vehicle system, the ONOS controller [48] controls the entire procedure. The network parameters used in proposed simulation approach design are shown in Table 2, i.e., vehicle velocity and RSU range of transmission, packet interval, etc. To vary the network density, we adjust the number of cars and their speed. In the proposed work, consider the vehicle and RSU transmission range are 200 and 500 meters, respectively.

Table 2 Simulation parameters

Parameters	Values
Packet size	512 bytes
No. of vehicles	200–500
Simulation time	500 s
Traffic type	CBR
Vehicle speed	5–25 m/s
Simulation area	2000 × 2500

4.2 Simulation Scenario

The metrics with a special emphasis on VANET applications' requirements are used to measure the successful routing efficiency using the intelligent vehicle system. Two edge controllers (EC) with one SDN controller in the control plane, the RSU are connected via high bandwidth internet to the centralized SDN controller to evaluate the proposed method using a vehicle topology. In the data plane, all presented as OVS, and we used vehicles and RSUs.

By adjusting the network density, the efficiency of the suggested vehicular intelligent system was assessed. The metrics are given as follows to evaluate the implementation.

Delivery Wait Period (DWP): The DWP is measured as the time utilized to effectively send number of data packets from sender to receiver. Basically the transmission delay time demonstrates the reliability of protocols for the efficient end-to-end transmission of data.

Packet delivery ratio (PDR): To measure effective routing reliability, this metric is established. PDR is considered as following in Eq. (1):

$$\text{Packet Delivery Ratio} = \frac{\text{Total Received Packets}}{\text{Total Send Packets}} \qquad (1)$$

4.3 Simulation Results

Packet Delivery Ratio (PDR): PDR decreases as the vehicle's speed increases shown in Fig. 6 and Table 3. With the car's high rate, the network's topology changes rapidly, increasing the packet loss ratio. It displays a relatively low PDR using AODV as well as GPSR due to the drawback of an AODV flooding system and a greedy GPSR forwarding approach. The flooding technique implemented by AODV takes up enormous device bandwidth, leaving less data transmission bandwidth and reducing the distribution ratio. When frugal mode traps in the local optimum, the greedy GPSR system turns to perimeter mode, increasing transmission delay. The proposed protocol ERS-SDN-IOV shows a higher performance compared to the AODV, GPSR, and SDGR.

PDR is also affected by the number of vehicle. The significant impact on PDR by the increment of the number of vehicle is shown in Fig. 7 and Table 3. It increases in a higher number of vehicle scenario since the less number of vehicle network cannot promise for necessary connectivity. The ERS-SDN-IOV approach's SDN framework offers a global view of traffic data for a routing approach to measure the shortest route with a higher number of vehicles. Thus, as opposed to SDGR, AODV, and GPSR, ESR-SDN-IOV has a greater PDR.

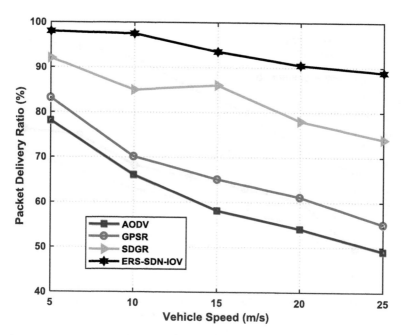

Fig. 6 Packet Delivery Ratio (PDR) versus vehicle speed (m/s)

Table 3 Performance Analysis of packet delivery ratio

	AODV	GPSR	SDGR	ERS-SDN-IOV
Speed	Packet delivery ratio vs. speed			
5	78	83	92	98
10	66	70	85	97.5
15	58	65	86	93.5
20	54	61	78	90.5
25	49	55	74	88.9
NO. of vehicle	Packet delivery ratio vs. vehicle			
200	58	62	65	75
250	66	70	72	79
300	68	72	78	85
350	70	74	80	89
400	72	76	85	94
450	75	81	88	96
500	78	82	92	97

Delivery Delay Time: Fig. 8 and Table 4 present a comparative study of delivery delay time. It shows that as the delivery time increases, so does the vehicle's speed. As the speed of the vehicle increases, then vehicle position changes very frequently. Thus the link and route stability became unstable and increase the packet transmission and transmission delay.

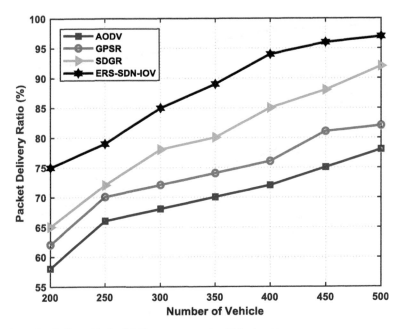

Fig. 7 Packet Delivery Ratio (PDR) versus several vehicle densities

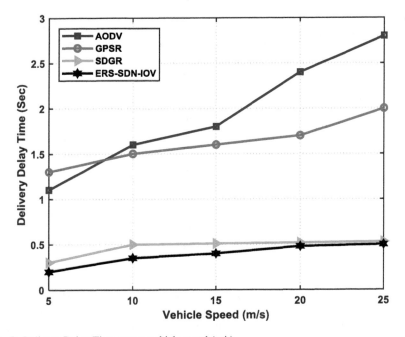

Fig. 8 Delivery Delay Time versus vehicle speed (m/s)

Table 4 Performance Analysis of delivery delay time

	AODV	GPSR	SDGR	ERS-SDN-IOV
Speed	Delivery Delay Time Vs. Speed			
5	1.1	1.3	0.3	0.2
10	1.6	1.5	0.5	0.35
15	1.8	1.6	0.51	0.4
20	2.4	1.7	0.52	0.48
25	2.8	2.0	0.53	0.5
NO. of vehicle	Delivery Delay Time Vs. Vehicle			
200	1.8	1.5	1.0	0.85
250	1.66	1.45	0.9	0.75
300	1.52	1.48	0.45	0.74
350	1.55	1.32	0.85	0.736
400	1.32	1.30	0.81	0.75
450	1.22	1.21	0.79	0.732
500	1.12	1.01	0.78	0.71

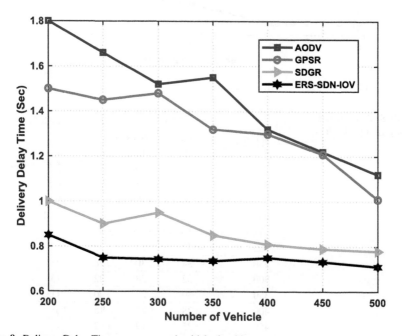

Fig. 9 Delivery Delay Time versus several vehicle densities

The higher number of vehicles in the network shows the less delivery delay time in Fig. 9 and Table 4. A higher number of vehicles can assure sufficient connectivity, route stability, and connection between the vehicle to its neighbor, which reduced the retransmission times as well as transmission delay. ERS-SDN-IOV hase lesser latency than AODV, SDGR, and GPSR, clearly shown in Fig. 9.

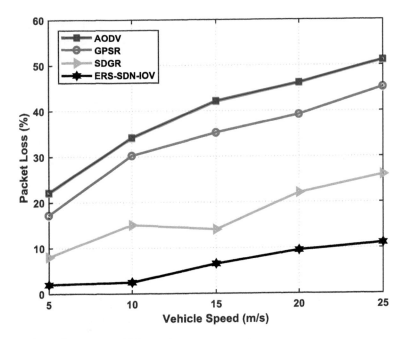

Fig. 10 Packet loss versus vehicle speed

Table 5 Performance Analysis of packet loss		AODV	GPSR	SDGR	ERS-SDN-IOV
	Speed	Packet Loss Vs. Speed			
	5	22	17	8	2
	10	34	30	15	2.5
	15	42	35	14	6.5
	20	46	39	22	9.5
	25	51	45	26	11.1
	NO. of vehicle	Packet Loss Vs. Vehicle			
	200	42	38	35	25
	250	34	30	28	21
	300	32	28	22	15
	350	30	26	20	11
	400	28	24	15	6
	450	25	19	12	4
	500	22	18	8	3

Packet Loss (PL): Packet loss increases as the vehicle's speed increases shown in Fig. 10 and Table 5. It demonstrates the high packet loss using AODV as well as GPSR. In the AODV 51% packet loss and GPSR show 41% packet loss at 25 m/s, which is relatively very high as compare to the proposed ERS-SDN-IOV 11.1% at 25 m/s. The proposed protocol ERS-SDN-IOV shows a higher performance and low packet loss as compared to the AODV, GPSR, and SDGR.

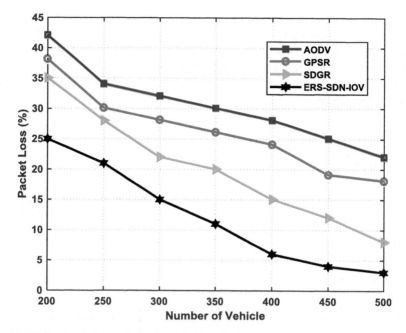

Fig. 11 Packet loss versus several vehicle densities

The variation of vehicle density is also affected the packet loss of the network. The significant impact on packet loss by the variation in number of vehicle is shown in Fig. 11 and Table 5. It decreases as number of vehicle in the scenario increases since the less number of vehicle network cannot promise for necessary connectivity. The ERS-SDN-IOV proposed approach shows the stable route and less number of packet loss as compare to other mentioned routing protocols SDGR, AODV, and GPSR.

5 Conclusion

This chapter introduces a software-defined network and vehicle internet location-based Efficient Routing System (ERS), which can provide a secure route by predicting the vehicle's location in the SDN-based IOV (SDN-IOV). To support location prediction and routing in SDN-IOV, designed a new proposed work by named of ERS-SDN-IOV. The simulation results of the proposed method (ERS-SDN-IOV) indicate better performance in both PDR and delay time compared to AODV, GPSR, and SDGR. The proposed (ERS-SDN-IOV) updated optimal forwarding route and shortest path with a higher number of vehicles leveraging the SDN controller's global view. The proposed (ERS-SDN-IOV) method reduces the local complexity and offers good connectivity as compare to existing methods. In vehicle load balancing, the congestion detection mechanism in SDGR plays a critical role, which effectively decreases packet loss ratio and transmission delay. So, for the SDGR, there is a much larger PDR and lower latency than both AODV and GPSR.

In future, we will test our proposed work with more QoS parameters with different network parameters such as pause time, network stop time, and high speed of the vehicle.

References

1. Abbas, M. T., Muhammad, A., & Song, W. C. (2020). SD-IOV: SDN enabled routing for internet of vehicles in road-aware approach. *Journal of Ambient Intelligence and Humanized Computing, 11*(3), 1265–1280.
2. Saleet, H., Langar, R., Naik, K., Boutaba, R., Nayak, A., & Goel, N. (2011). Intersection-based geographical routing protocol for VANETs: A proposal and analysis. *IEEE Transactions on Vehicular Technology, 60*(9), 4560–4574.
3. Abbasi, I. A., Nazir, B., Abbasi, A., Bilal, S. M., & Madani, S. A. (2014). A traffic flow-oriented routing protocol for VANETs. *EURASIP Journal on Wireless Communications and Networking, 2014*(1), 1–14.
4. Salkuyeh, M. A., & Abolhassani, B. (2016). An adaptive multipath geographic routing for video transmission in urban VANETs. *IEEE Transactions on Intelligent Transportation Systems, 17*(10), 2822–2831.
5. Salkuyeh, M. A., & Abolhassani, B. (2016). An adaptive multipath geographic routing for video transmission in urban VANETs. *IEEE Transactions on Intelligent Transportation Systems, 17*(10), 2822–2831.
6. Yaqoob, S., Ullah, A., Akbar, M., Imran, M., & Shoaib, M. (2019). Congestion avoidance through fog computing in internet of vehicles. *Journal of Ambient Intelligence and Humanized Computing, 10*(10), 3863–3877.
7. Hussain, N, & Rani, P. (2020). *Comparative studied based on attack resilient and efficient protocol with intrusion detection system based on deep neural network for vehicular system security.* Distributed Artificial Intelligence: A Modern Approach (p. 217).
8. Hussain, N., Singh, A., & Shukla, P. K. (2016). In depth analysis of attacks & countermeasures in vehicular ad hoc network. *International Journal of Software Engineering and Its Applications, 10*(12), 329–368.
9. Weng, J. S., Weng, J., Zhang, Y., Luo, W., & Lan, W. (2018). BENBI: Scalable and dynamic access control on the northbound interface of SDN-based VANET. *IEEE Transactions on Vehicular Technology, 68*(1), 822–831.
10. Zhu, M., Cao, J., Pang, D., He, Z., & Xu, M. (2015, August). SDN-based routing for efficient message propagation in VANET. In *International conference on wireless algorithms, systems, and applications* (pp. 788–797). Springer.
11. Bhatia, J., Dave, R., Bhayani, H., Tanwar, S., & Nayyar, A. (2020). SDN-based real-time urban traffic analysis in VANET environment. *Computer Communications, 149*, 162–175.
12. Shafiq, H., Rehman, R. A., & Kim, B. S. (2018). Services and security threats in SDN based VANETs: A survey. *Wireless Communications and Mobile Computing, 2018*, 8631851.
13. Kai, K., Cong, W., & Tao, L. (2016). Fog computing for vehicular ad-hoc networks: paradigms, scenarios, and issues. *The journal of China Universities of Posts and Telecommunications, 23*(2), 56–96.
14. Pereira, J., Ricardo, L., Luís, M., Senna, C., & Sargento, S. (2019). Assessing the reliability of fog computing for smart mobility applications in VANETs. *Future Generation Computer Systems, 94*, 317–332.

15. Mukherjee, M., Shu, L., & Wang, D. (2018). Survey of fog computing: Fundamental, network applications, and research challenges. *IEEE Communications Surveys & Tutorials, 20*(3), 1826–1857.
16. Bylykbashi, K., Qafzezi, E., Ikeda, M., Matsuo, K., & Barolli, L. (2020). Fuzzy-based driver monitoring system (FDMS): Implementation of two intelligent FDMSs and a testbed for safe driving in VANETs. *Future Generation Computer Systems, 105*, 665–674.
17. Garg, S., Singh, A., Kaur, K., Aujla, G. S., Batra, S., Kumar, N., & Obaidat, M. S. (2019). Edge computing-based security framework for big data analytics in VANETs. *IEEE Network, 33*(2), 72–81.
18. Anwer, M. S., & Guy, C. (2014). A survey of VANET technologies. *Journal of Emerging Trends in Computing and Information Sciences, 5*(9), 661–671.
19. Truong, N.B., Lee, G.M. and Ghamri-Doudane, Y., 2015, May. Software defined networking-based vehicular adhoc network with fog computing. In *2015 IFIP/IEEE International Symposium on Integrated Network Management (IM)* (pp. 1202-1207). IEEE.
20. Ydenberg, A., Heir, N., & Gill, B. (2018, January). Security, SDN, and VANET technology of driver-less cars. In *2018 IEEE 8th Annual Computing and Communication Workshop and Conference (CCWC)* (pp. 313–316). IEEE.
21. Jaballah, W. B., Conti, M., & Lal, C. (2020). Security and design requirements for software-defined VANETs. *Computer Networks, 169*, 107099.
22. Arif, M., Wang, G., Geman, O., Balas, V. E., Tao, P., Brezulianu, A., & Chen, J. (2020). Sdn-based vanets, security attacks, applications, and challenges. *Applied Sciences, 10*(9), 3217.
23. Li, T., Chen, J., & Fu, H. (2019, April). Application scenarios based on SDN: an overview. In *Journal of Physics: Conference Series* (Vol. 1187, No. 5, p. 052067). IOP Publishing.
24. Abuashour, A., & Kadoch, M. (2017, August). An intersection dynamic VANET routing protocol for a grid scenario. In *2017 IEEE 5th International Conference on Future Internet of Things and Cloud (FiCloud)* (pp. 25–31). IEEE.
25. Ku, I., Lu, Y., Gerla, M., Gomes, R. L., Ongaro, F., & Cerqueira, E. (2014, June). Towards software-defined VANET: Architecture and services. In *2014 13th annual Mediterranean ad hoc networking workshop (MED-HOC-NET)* (pp. 103–110). IEEE.
26. Ji, X., Yu, H., Fan, G., & Fu, W. (2016, December). SDGR: An SDN-based geographic routing protocol for VANET. In *2016 IEEE International Conference on Internet of Things (iThings) and IEEE Green Computing and Communications (GreenCom) and IEEE Cyber, Physical and Social Computing (CPSCom) and IEEE Smart Data (SmartData)* (pp. 276–281). IEEE.
27. Zhu, M., Cao, J., Pang, D., He, Z., & Xu, M. (2015, August). SDN-based routing for efficient message propagation in VANET. In *International conference on wireless algorithms, systems, and applications* (pp. 788–797). Springer.
28. Berde, P., Gerola, M., Hart, J., Higuchi, Y., Kobayashi, M., Koide, T., Lantz, B., O'Connor, B., Snow, G., & Parulkar, G. (2014, August). ONOS: towards an open, distributed SDN OS. In *Proceedings of the third workshop on Hot topics in software defined networking* (pp. 1–6).
29. Venkatramana, D. K. N., Srikantaiah, S. B., & Moodabidri, J. (2017). SCGRP: SDN-enabled connectivity-aware geographical routing protocol of VANETs for urban environment. *IET Networks, 6*(5), 102–111.
30. Jibran, M. A., Abbas, M. T., Rafiq, A., & Song, W. C. (2020). Position prediction for routing in software defined internet of vehicles [J]. *Journal of Communications, 2020*, 157–163.
31. Alvear, O., Calafate, C. T., Cano, J. C., & Manzoni, P. (2018). Crowdsensing in smart cities: Overview, platforms, and environment sensing issues. *Sensors, 18*(2), 460.
32. Bisio, I., Lavagetto, F., Marchese, M., & Sciarrone, A. (2015). Smartphone-centric ambient assisted living platform for patients suffering from co-morbidities monitoring. *IEEE Communications Magazine, 53*(1), 34–41.
33. Shafik, W., Matinkhah, S. M., & Ghasemzadeh, M. (2020). Internet of things-based energy management, challenges, and solutions in smart cities. *Journal of Communications Technology, Electronics and Computer Science, 27*, 1–11.

34. Ang, L. M., Seng, K. P., Ijemaru, G. K., & Zungeru, A. M. (2018). Deployment of IOV for smart cities: applications, architecture, and challenges. *IEEE Access, 7,* 6473–6492.
35. Ejaz, W., & Anpalagan, A. (2019). Internet of Things enabled electric vehicles in smart cities. In *Internet of Things for smart cities* (pp. 39–46). Springer.
36. Shah, S. S., Ali, M., Malik, A. W., Khan, M. A., & Ravana, S. D. (2019). vFog: A vehicle-assisted computing framework for delay-sensitive applications in smart cities. *IEEE Access, 7,* 34900–34909.
37. Kang, J., Yu, R., Huang, X., & Zhang, Y. (2017). Privacy-preserved pseudonym scheme for fog computing supported internet of vehicles. *IEEE Transactions on Intelligent Transportation Systems, 19*(8), 2627–2637.
38. Gil, D., Ferrández, A., Mora-Mora, H., & Peral, J. (2016). Internet of things: A review of surveys based on context aware intelligent services. *Sensors, 16*(7), 1069.
39. He, X., Wang, K., Huang, H., & Liu, B. (2018). QoE-driven big data architecture for smart city. *IEEE Communications Magazine, 56*(2), 88–93.
40. Li, R., Zhao, Z., Zhou, X., Ding, G., Chen, Y., Wang, Z., & Zhang, H. (2017). Intelligent 5G: When cellular networks meet artificial intelligence. *IEEE Wireless Communications, 24*(5), 175–183.
41. Longo, A., Zappatore, M., Bochicchio, M., & Navathe, S. B. (2017). Crowd-sourced data collection for urban monitoring via mobile sensors. *ACM Transactions on Internet Technology (TOIT), 18*(1), 1–21.
42. Corradi, A., Foschini, L., Gioia, L., & Ianniello, R. (2016, December). Leveraging communities to boost participation and data collection in mobile crowd sensing. In *2016 IEEE Global Communications Conference (GLOBECOM)* (pp. 1–6). IEEE.
43. Panichpapiboon, S., & Leakkaw, P. (2017). Traffic density estimation: A mobile sensing approach. *IEEE Communications Magazine, 55*(12), 126–131.
44. Abdellah, A. R., Muthanna, A., & Koucheryavy, A. (2019). Robust estimation of VANET performance-based robust neural networks learning. In *Internet of Things, smart spaces, and next generation networks and systems* (pp. 402–414). Springer.
45. Ovasapyan, T. D., Moskvin, D. A., & Kalinin, M. O. (2018). Using neural networks to detect internal intruders in VANETs. *Automatic Control and Computer Sciences, 52*(8), 954–958.
46. https://www.nsnam.org/
47. https://sumo.dlr.de/docs/Tutorials/Trace_File_Generation.html
48. https://opennetworking.org/onos/

An Enhanced Cloud-IoMT-based and Machine Learning for Effective COVID-19 Diagnosis System

Joseph Bamidele Awotunde, Sunday Adeola Ajagbe, Ifedotun Roseline Idowu, and Juliana Ngozi Ndunagu

1 Introduction

The emerging of coronavirus, a family of novel severe contagious respiratory syndrome called (COVID-19) has caused the greatest public health challenge globally, after the pandemic of the influenza outbreak of 1918. According to the World Health Organization, as of 5:02 pm CEST, 23rd November 2020, 58,425,681 confirmed, and 1,385,218 (2.4%) death cases globally. Globally, people spend much of their time indoor to contain or avoid people infected with the virus. There are been various research towards finding a lasting solution to the COVID-19 outbreak that is treating the whole world.

Around ten billion people from various cities and towns are facing serious problems with self-quarantine themselves at home with lockdown measures. In recent time, the medical tools and supplies are of increased due to the pandemic, and almost all the existing one vitally need necessary replenishment. To seek medical help, potential citizens have to leave their homes, and that post-real risks

---

J. B. Awotunde (✉)
Department of Computer Science, University of Ilorin, Ilorin, Nigeria
e-mail: awotunde.jb@unilorin.edu.ng

S. A. Ajagbe
Department of Computer Engineering, Ladoke Akintola University of Technology LAUTECH, Ogbomoso, Nigeria
e-mail: saajagbe@pgschool.lautech.edu.ng

I. R. Idowu
Department of Computer Science, Kwara State University, KWASU, Ilorin, Nigeria
e-mail: ifedotun.idowu18@kwasu.edu.ng

J. N. Ndunagu
Department of Computer Science, National Open University of Nigeria, Abuja, Nigeria
e-mail: jndunagu@noun.edu.ng

© The Author(s), under exclusive license to Springer Nature Switzerland AG 2021
F. Al-Turjman et al. (eds.), *Intelligence of Things: AI-IoT Based Critical-Applications and Innovations*,
https://doi.org/10.1007/978-3-030-82800-4_3

for the unaffected citizens and a hole on the efforts of isolation and quarantine put in place to contain the outbreak globally [1–3]. Also, the shortage of medical personnel, lack of proper medical equipment, and the shortage of isolation clinics have prompted the policymakers to encourage those with suspected symptoms, or mild stages to stay at home. Hence, there is an urgent need for a home-based diagnostic test to serve as an alternative method that will be cost-effective, flexible, and can deliver invaluable solutions for the self-isolated patient [4].

In today's telemedicine, in diagnosis, prediction, remote monitoring of patient, treatment, and therapy, the Internet of Things (IoT) has increasingly play prominent roles, thus experienced rapid acceptance in the healthcare sectors with an emphasis on designing smart applications. The Internet of Medical Things (IoMT) provides a forum for devices and sensors to interact smoothly within a smart world and facilitate easy exchange of data and information through the Internet. Modern implementations of several wireless equipment positions Cloud-IoMT-based has an advanced technology by taking delivery of the full prospects that digital technology offers. In healthcare sectors, IoT-based system has drawn convincing ground in various fields, especially in the healthcare sectors. This new revolution reshapes the traditional healthcare systems by integrating new technological prospects that reduced the healthcare cost economically and socially. The healthcare systems are more becoming personalized systems when compared with the conventional healthcare systems, where patients can be monitored, diagnosed, and treated effortlessly. Wearable body sensor network has transformed the power to change people's lifestyle with abundant technologies especially in areas of healthcare, entertainment, transportation, business, and emergency services control. The integration of wireless sensors with simulation and intelligent systems research has developed an interdisciplinary definition of ambient intelligence to address the obstacles faced in everyday lives.

The IoMT is detailed of the IoT-based system for the healthcare system [5, 6]. The IoMT could be used to help patients get proper medical care at home when applied during this epidemic, and the healthcare policymakers and government can make use of the robust database created for COVID-19 outbreak management. Diagnostic and healthcare devices such as thermometers, smart helmet, smart wristwatch, medications, protective masks, and monitoring infection kits may be purchased for people with moderate symptoms. The patients can upload their health status periodically over the Internet based on the wireless network to the clinical cloud storage, and their data could be forwarded to nearest clinics or health center hospitals, and the Center for Disease Control (CDC) [7].

Subsequently, a medical expert will provide online health consultations based on the health status of each patient and if necessary, the policymakers and healthcare experts assign facilities and designate quarantine stations to the affected person. People may dynamically monitor their clinical diagnosis and obtain adequate medical needs using the IoMT platform without virus transmission to others. Hence, minimize the costs, alleviate the shortages of medical equipment, and provide a systemic database that could be used by physicians to track the spread of COVID-19 effectively, the supplies of relative tools become easier, and enforce emergency strategies. The current pandemic we are fighting is increasingly suffocating the healthcare sector towards its ending levels; the hospitals and clinics are filled with reported suspected cases pending the evidence of the diagnosis. As the demands are rising, the shortage of medical diagnostic equipment and supplies is increasing, and with increases in patients who need care without adequate tools for complementing the upward increase in the number of patients in the hospital, and the place mission-critical healthcare staff at higher risk on the front lines. A powerful and supportive medical system is needed to save the lives of many during the COVID-19 outbreak and to mitigating the crisis bring this period.

A more robust medical framework to fight against the COVI-19 pandemic is needed. To alleviate the diagnostic and monitoring problems, an IoMT-based system is needed, and this will help in enforcing stay-at-home protocols and limiting the clinical resources required. This will support the appropriate distribution of equipment and supplies by government and private donors to clinics and various hospitals, and the approach would provide information on healthcare facilities to establish effective patient care. To save countless lives, this combined strategy can be very helpful, and also safeguard strained economies and build a blueprint to tackle future threats more effectively (Fig. 1).

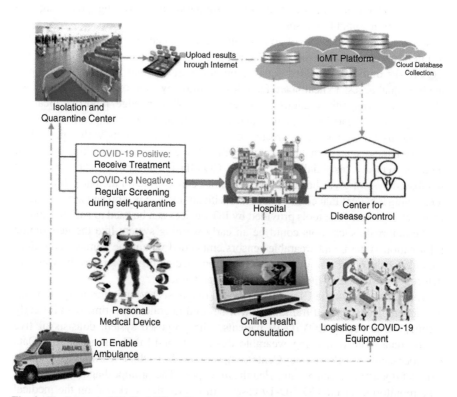

Fig. 1 The Internet of Medical Things Ecosystem for Combating COVID-19 Outbreak

In an attempt to prevent the circulation of the sickness, several countries were extending the lockdown and affecting millions of people. Initially, experts raised questions about the feasibility of this initiative and cautioned that certain countries were at risk of repeating a SARS-like epidemic [8, 9]. By this time, the pandemic appears to be under control in most countries, although there is still criticism of the utilization of what others have called "draconian" steps to stifle its circulation. The globe is presently trying to monitor the exceptional virus circulation that involves the greatest sum of indispositions plus deaths. As there is no such thing as definite medical care for coronaviruses and attempts to control the circulation has yet remained unsuccessful [10, 11], there is a crucial necessity for worldwide investigation of persons with intense COVID-19 contagion.

The emerging of new trends in technologies contributed to the commencement of the IoT and is acquiring worldwide concentration as well as becoming obtainable for monitoring, diagnosing, forecasting, and preventing arising communicable ailments. IoT in the medical organization is advantageous and enabled suitable controlling of COVID-19 persons by using interrelated wearable sensors and networks. IoT is an evolving area of investigation within infectious disease epidemiology. However, the augmented dangers of communicable ailment transmitted over worldwide integration and the pervasive obtainability of smart types of machinery, including interrelatedness of the world require its utilization for monitoring, averting, predicting, and managing transmittable viruses.

IoT is an innovative way of combining healthcare gadgets and their applications to interact with human resources. The adoption of an IoT-based system during the COVID-19 outbreak provided equal rights for both rich and the poor populace in having equal access to healthcare facilities without any form of preference. Various cloud-based IoT administrations are the exchange of knowledge, report verification, investigation, diagnosis, treatment, and patient monitoring among other services provided by the system. This creates rewarding various patients with a prevailing treatment and diagnosis that are more fulfilling and creates a new working system of medical services, especially during the COVID-19 pandemic. The use of an IoT-based system gives health workers to fully concentrate on the patient by easily identifying a patient that comes in contact with an infected person and move them to an isolation center. The tools provided by IoT devices can be used to curb the spread of the outbreak, such tools could be an early warning system like the geographic information system, and wearable sensors embedded within the human body. Sensors like temperature and other signs might be used at airports around the world to detect people infected with the COVID-19 pandemic.

Therefore, a real-time Cloud-IoMT-based diagnosis and monitoring system was developed. The proposed framework will be used to collect real-time data for early identify of suspected COVID-19 patients. The proposed system consists of five layers: data collection using wearable devices, Cloud-IoMT infrastructure, Data Preprocessing, and Processing Center empower with machine learning algorithms, Isolation/Quarantine center, and Healthcare expert. The prompt diagnosis and real-time monitoring of the COVID-19 cases will reduce the workload on the medical personnel for the control and prevention of the COVID-19.

The proposed method has the following contributions:

- The study uses ML algorithms for the diagnosis, and prediction based on COVID-19 dataset, this study is the first to use four ML to the best of authors' knowledge.
- The significant of cloud, IoMT, and ML are highlighted in smart healthcare system to solve the real-time diagnosis, and prediction during the COVID-19 pandemic.
- The study validates the four machine learning: Extra Trees, Random Forest (RF), XGBoost, and Light Gradient Boosting Machine (LGBM) on COVID-19 diagnosis results based on normalized datasets.
- The performance evaluation of the proposed method was carried out using accuracy, precision, F1-score, AUC, and recall, and the method was compared with state-of-the-art existing methods.

The chapter is organized as follows. Section 2 explains the applications of cloud computing and IoMT to combat the COVID-19 pandemic. Section 3 presents related work. Section 4 presents the Cloud-IoMT-based diagnosis and monitoring framework for the fighting COVID-19 outbreak. Section 5 discusses the practical applicability of the proposed model, while Sect. 6 presents the results and discussion, and Sect. 7 presents conclusion and future works for the realization of efficient uses of Cloud-IoT-based in fighting the COVID-19 pandemic.

2 Application of Cloud Computing and IoMT to Combat COVID-19 Pandemic

The healthcare sector is changing rapidly in advanced nations as expectancy increased significantly during the 1990s [12]. In these regions of the world, infectious diseases are also putting significant pressure on health care systems [13]. Over the twentieth century, life expectancy in industrialized nations rose by around 30 years. The number of elderly people has increased rapidly as a result [12]. In addition, due to a shortage of resources, chronic condition globalization has placed new demands on health care systems in other countries [13]. There are great obstacles to continuing to increase infectious diseases and aging populations, as health systems have to deal with a multitude of diseases and treatment options, but also with an increasing number of patients. Effective techniques have been demonstrated in in-house telemedicine systems to prevent overloading health facilities and reducing healthcare costs [14].

Telemedicine systems, such as mobile tele-monitoring and trauma rehabilitation, are highly diversified and usually configured to react to a single therapeutic purpose [15]. This characteristic of telemedicine capabilities makes health systems cost-effective and overloaded, but represents a vulnerability as the number of patients and the variety of diseases increase. The need for improved genericity and reliability can be addressed by the IoT. The IoT, of course, combines the reliability and security of conventional medical devices as well as traditional IoT dynamics, genericity, and scalability capabilities. Through handling multiple sensors deployed for millions of patients, it recognizes the ability to solve the aging problem and terminal diseases, as well as being large enough to deal with multiple diseases requiring precise, varied control and action requirements.

The Internet of Things (IoT) in transmittable sickness epidemiology is an evolving area, but the omnipresent proliferation of smart technology and amplified threats of transmittable sickness conveyed via global integration and worldwide interconnectivity demand its use to anticipate, deter, and monitor evolving COVID-19 pandemics [16, 17]. Web-based monitoring platforms and strategies for disease intelligence have recently appeared in many countries [16, 17] to promote risk management and prompt identification of outbreaks, but there is a shortage of systematic use of available technologies. IoT-Implemented Medical Care Surveillance in a worldwide health care system would offer local health authorities the ability to strengthen efforts to identify, control, and avoid infectious diseases [18].

In the field of healthcare, IoT devices such as satellites could have diverse technologies, blood pressure control, glycemic control, and endovascular tablets, for example [19, 20]. The integration of devices, actuators, and other mobile technology equipment will turn the functioning of the medical industry, especially during any outbreak of disease [21, 22], into a combined smart healthcare device framework that receives information subsequently offered to healthcare IT systems through web communications systems [3]. Currently, 3.7 million therapeutic devices are now in use, and are then attached to and tracked by various sections of the body to record medical resolutions [23, 24].

These IoT devices connect to public clouds such as Google Virtualization, Microsoft Azure Data center, Amazon Web Services, or other storage and analytics information collected by any other custom web services. In addition, IoT programs provide remote medical observation for individuals with chronic and long-term circumstances. These systems can monitor any orders for patient treatment and track the status of patients in hospitals and clinics who are put on wearable healthcare devices. It is possible to send the obtained medical information to their care provider. Transfusion engines that connect to observational control panels and hospital wards with sensors that monitor the vital signs of patients are medical devices that can be incorporated or implemented as IoT technologies. Through the use of objects, i.e., "Smart" products, that use multiple sensors and actuators that are fully prepared to learn in their context, and through the use of embedded communication networks to connect with any potential alternatives, the IoT obtains its maximum capabilities [25].

Cloud computing will play an important role in absorbing healthcare transformation expenses, optimizing assets, and bringing the new age of technology to life. Emerging policies are targeted at obtaining data at anytime, anywhere that can be achieved by transferring health data to the cloud. This contemporary delivery model will make healthcare more productive and effective and lowering the price of innovation expenditures [26], but it also presents some obstacles due to issues regarding the security of sensitive health information and compliance with a specific criterion such as HIPAA. Healthcare providers, taking into account these security and privacy concerns, can unquestionably reap the benefits of cloud computing technologies and provide substantial benefits, such as assisting to improve the quality of service for patients and reducing healthcare spending [27].

Cloud computing's critical features are (1) self-service on-demand REnumber accordingly, (2) wide network access, (3) rapid elasticity, and (4) calculated services. In complex resources, clouds offer advantages such as processing power or storage capacities, ubiquitous access to resources from anywhere at any time, and high resource versatility and scalability. In several business fields, these advantages have been the reason for the growing adoption of cloud computing. This principle has also evidently been adopted in the area of healthcare in recent years. At least in the mainstream literature and is provided by healthcare IT firms, but even in the scientific literature cloud computing for healthcare applications is gaining interest, a continuously growing number of papers and publications appears.

The capacity to share data between different systems would be one of the main advantages of cloud computing. This capacity is something that IT urgently needs for healthcare. Cloud computing, for example, can enable health care professionals to share data such as EHR, doctor's references, medications, insurance data, research reports stored via various information systems. In the radiological market, where many organizations have switched to the cloud to minimize their computing costs, and promote the sharing of pictures, this is already happening [27]. Cloud computing has provided clinics, hospitals, insurance providers, pharmacies, and other healthcare companies the ability to agree to cooperate and exchange healthcare data to provide improved service quality and minimize costs. Looking at the developments in the industry, it seems that once all the obstacles it presents are resolved, cloud-based systems will eventually become the standard in healthcare.

3 Related Works

A smart healthcare hospital was proposed in urban areas by Rath and Pattanayak [28] using IoT devices. In the VANET region, a timely healthcare system for patients was addressed with issues like hygiene and safety of patients. The NS2 and NetSim Simulators were used to test the proposed method [28]. Darwish et al., [29] suggested, based on the related literature, a Cloud IoT-Health model that combines cloud computing with IoT in the health sector. As well as emerging developments in CloudIoT-Health, the paper addressed the complexities of integration. Three layers identify these challenges: technology, connectivity and networking, and knowledge. Zhong [30] studied tracking physical activities of college students within the school premises. The author concentrated on a model of bodily movement identification and tracking (PARM). Three algorithms were used for the comparison of the results of the proposed system (decision tree, neural networks, and SVM).

Din and Paul [31] proposed an IoT-based system for healthcare management and monitoring scheme. The framework was divided into three layers: (i) The data capture through the medical sensors, (ii) the Hadoop platform for data processing, (iii) the device's center. An energy harvesting method with piezoelectric tools connected to the human body was used by the proposed, due to the limited ability of batteries to power the sensors. Otoom et al., [32] built a blood sugar regulation and real-time monitoring system. To assess the insulin dosage, Markov-based and the ARIMA replicas were used. In [34], various ML-based models were used for cardiovascular disease detection using the capture data from IoT-based system.

To increase safety in the outdoors, a hybrid IoT-based healthcare monitoring and protection was developed by Wu et al., [35]. The system was divided into two parts: the data capture layer and the cloud database storage layer, and the collected data was process using ML-based model. The wearable devices were used to gather patients' health symptoms and safety signs from the environment. An Industrial IoT (HealthIoT) a real-time health monitoring system was implemented to negate death conditions by Hossain and Muhammad [36]. This device has considerable capacity for evaluating patient health data. This IoT system for healthcare uses medical equipment and sensors to capture patient data. Also, safety measures such as watermarking and signal enhancement have been introduced into this system to prevent identity fraud or clinical mistakes by health professionals. In 2016, Gope and Hwang [37] identified a new technology known as the body sensor network (BSN) based on the advances of IoT medical devices.

The patient can be tracked using various tiny-powdered and light-weight sensor networks in this system. Besides, the safety criteria for the implementation of the BSN-healthcare system have also been considered in this context. In 2015, Verma and Sood [38] addressed the context of IoT along with its implementation from the viewpoint of u-healthcare. The authors proposed an ideological IoT structure for u-healthcare. The heterogeneity problem of the data format in the IoT platform was solved by Xu et al., [38] in 2014 using the semantic data model. Besides, the resource-based data accessing system (UDA-IoT) is designed to ubiquitously process IoT data. Also, to demonstrate the collection, integration, and interoperation of IoT data, an IoT-based framework for handling medical emergencies was presented. The new methods and algorithms for analyzing data obtained from wearable sensors in the health monitoring setting were clarified in 2013, Banaee et al., [39]. Continuous time-series measurements obtained from wearable sensors have been applied to data mining tasks such as anomaly detection, prediction, and decision making. The methodologies for designing m-health-based apps were discussed in 2014, Zhang et al., [40]: namely website builder and application builder to remotely track patients using IoT-based healthcare medical system. Web-based applications have been created to offer patients' health information to respondents (doctors) outside a medical environment. Also, IoT-based health monitoring was used by these authors to assess adverse health effects, including alcohol consumption and therapeutic impacts of medical interventions [41–44].

The authors in [45, 46] conducted a survey on the use of artificial intelligence (AI) and machine learning (ML) algorithms during COVID-19 pandemic. These techniques were categorized by their work into many groups, including the use of IoT. Rao and Vazquez [47] suggested the use of algorithms for machine learning to classify potential cases of COVID-19. A web survey through a smartphone was used to collect data that was used for the learning of the proposed framework. Allam and Jones [48] addressed the need to establish uniform protocols, inspired by the outbreak of COVID-19, to exchange knowledge between smart cities during pandemics. For example, to identify the potential COVID-19 pandemic, IoT-based devices were used with the help of AL models based on the capture data using thermal cameras installed in smart cities. Fatima et al., [49] used fuzzy inference engine to classify COVID-19 cases based on the capture data using IoT-based devices. The strategy is based on a fuzzy inference system. Peeri et al., [50] used the available works like MERS, SARs, and COVID-19 cases to perform a comparison with IoT-based data capture by the used devices to reduce the spread of COVID-19 infection. Sharma et al., [51] proposed drone-based dynamic model and control techniques, the model was efficient, accurate, and effective for healthcare system compared to the existing models.

4 The Proposed Cloud-IoMT-Based Architecture for Early Diagnosis of COVID-19 Outbreak

The proposed framework was depicting and discusses in this section. The proposed architecture could be used to diagnose a potential COVID-19 case in real time, and most especially the proposed framework can be used for the monitoring and predicting treatment for the confirmed patients with a better understanding of the nature of the COVID-19 outbreak. Figure 3 displayed the architecture of the Cloud-IoT-based Diagnosis and Monitoring for COVID-19 pandemic. The proposed system contains five distinct layers: The data capture collection (Symptom), Isolation/Quarantine layer, Data Preprocessing, and Analysis layer, and the Application layer for Health Physicians. The layers were interconnected through Cloud Infrastructure.

The ML models play a major role in the decision-making process, even when the data volume is very large [52–54]. The method of implementing data processing techniques for particular fields requires specifying types of data such as velocity, variety, and volume. Normal data analysis modeling involves the model of the neural network, the model of classification and the process of clustering, and the implementation of efficient algorithms as well [54–56]. Data can be generated from different sources with specific types of data, and it is also important for the development of methods capable of handling data characteristics. The proposed Cloud-IoMT-based architecture was displayed in Fig. 2. The framework has functional components: Data capture and collection center, the isolation/Quarantine layer, the Data Preprocessing, and Processing with the help of ML, Health Physicians alert center, all the components were interconnected through a wireless gateway and Cloud Infrastructure [57].

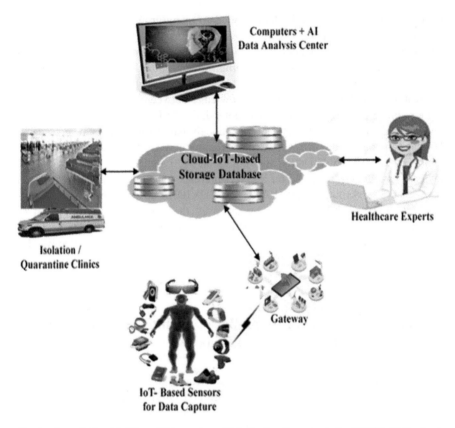

Fig. 2 General Cloud-IoT-based Diagnosis and Monitoring Framework for COVID-19 Pandemic

4.1 Data Capture and Collection of Symptoms Layer

The layer was used to capture and collect the COVID-19 symptoms like Fever, Sore throat, Shortness of Breath, Fatigue, and Cough been the most relevant symptoms [58, 59]. The layer contains wireless sensors, which can be move around by the carriers of the IoT-based devices. The sensors are divided into embedded [60] and body wearable sensors (WSs) [61, 62].

In any situation, a person can have several wearable sensors, and very possible for a patient to have more than one WSs. But experts determine the total number of WSs and this also depends on the required information. To detect these symptoms, several biosensors are available. For example, for the detection of fever [63, 64], audio-based sensors can be used for aerodynamic systems, and for acoustic, temperature-based sensors are very appropriate [65], for fatigue diagnosis, classifications based on ages of patients, heart rate, motion-based sensors, and cough can be used for this type of diseases [66]. For the diagnosis of Sore Throat, an image-based classification can be used [67]. Finally, it is possible to use oxygen-based sensors to detect shortness of breath [68].

4.2 Isolation/Quarantine Layer

The components are used to isolate and quarantine an infected person, this can be obtained from the gathered data from users in a healthcare facility. The time-series of the COVID-19 outbreak are containing in each record capture from the wearable sensors like history of any diseases in a family, travel history, age, gender, and other relevant information that may be very useful. The treatment response for each case would ultimately also include each record.

4.3 Computers and AI Data Analysis Center

The computer and AI data analysis model generate a real-time prediction, diagnosis, and monitoring events and produce the required reports needed by experts. To enhance the IoT-based devices' results that are characterized by low computing skill, the layer provides a high computing ability. The real-time results from this layer can be used by the physicians and medical experts using various predictor models that report COVID-19 suspicious cases, thus help doctors and experts in prompt decision making and advise accordingly. The AI data analysis (i) used the capture data to generate real-time diagnosis and monitor undesired events and signs related to the outbreak, and (ii) help to determine the positions of IoT-based devices in the 3D environment and the total number of IoT-based devices.

4.4 Healthcare Experts

The healthcare experts will be able to monitor suspected cases whose symptom data captured in real-time, and suggest a potential septicity based on the proposed model centered on machine learning. The clinical examination is required to validate the results by the reasons to follow up. This enables the reported cases to be separated and sufficient health care provided.

4.5 Cloud-IoT-based Storage Database

This component is responsible for communicating with IoT-based devices and patient devices used to identify biological and physiological signs of a person and conduct primary data analysis. The result of this section is a description of the circumstances of patients sent to healthcare experts. The interconnected system using the Internet allows (1) each user to upload capture symptoms, (2) keeps personal health records of each user, (3) links prediction results, (4) sends advice to the physician, and (5) stores details. Also, the framework can respond to signs of irregularity when it is identified, e.g., submitting a request for assistance (e.g., demand for an assistant care provider) or an urgent request (e.g., call for an ambulance) when an urgent condition is detected.

4.6 Gateway

This component is responsible for communicating with patient devices used to identify symptoms of patients and conduct primary data analysis. The result of this section is a description of the circumstances of patients sent to healthcare. Also, the framework can respond to signs of irregularity when it is identified, e.g., submitting a request for assistance (e.g., demand for an assistant care provider) or an urgent request (e.g., call for an ambulance) when an urgent condition is detected.

5 Applicability of the Proposed Model

The applicability of the proposed model in this chapter basically entails: (1) preprocessing of the data captured using Io-MT device and (2). diagnosis models, parameters, and metrics.

5.1 Preprocessing

Presenting the study as a classification problem: Let $S = \{(x_1, y_1), (x_2, y_2), \cdots (x_n, y_n)\}$ be the set of training instances of dimension d. $Y = \{y_1, y_2 \cdots, y_n\}$ be the set of labels (COVID-19, PTB, and Normal) where x_i is a feature with corresponding y_i label. The initial step taken in the image classification model is the extraction of features. This is pertinent when the features extracted which is also the input data is extremely large and difficult to process in its raw form. The selection of important features will resolve this problem, this was done according to [69].

5.2 Diagnosis Models, Parameters, and Metrics

This section describes the models, their parameters, and metrics used for the multi-class classification experiment. The task was performed with four learners: LGBM, Extra Trees, RF, XGBoost. Table 1 presents the parameters of the model used in this study. A confusion matrix is a table representing the prediction performance of a model. The row and column represent the predicted and the actual class, respectively, as shown in Table 2. The formula for metrics computed from the confusion matrix is presented in Eqs. (1)–(5).

$$\text{Accuracy} = \frac{TP + TN}{TP + TN + FP + FN} \quad (1)$$

$$\text{Precision} = \frac{TP}{TP + FP} \quad (2)$$

$$\text{Recall (TP Rate)} = \frac{TP}{TP + FN} \quad (3)$$

$$\text{FP Rate)} = \frac{FP}{FP + TN} \quad (4)$$

$$F1 - \text{Score} = 2 \times \frac{\text{Precision} \times \text{Recall}}{\text{Precision} + \text{Recall}} \quad (5)$$

6 Results and Discussion

This section presents the results and discusses the discoveries in the study. The experiment was performed with Jupyter notebook with sklearn [70] libraries on the Anaconda platform. All experiments were performed on an Intel® core™ i5–7200 CPU @ 2.50GHz to 2.70 GHz Pentium Windows computer with 8GB RAM. The

Table 1 Model parameters applied in this study

Models	Parameters
LGBM	n_estimators=100, random_state =10
Extra Trees	random_state=10
RF	n _ estimators = 700, max _ depth = 3
XGBoost	Learning _ rate = 0.05, max _ depth = 40, max _features = 1.0, min _ samples _ leaf = 4, n _ estimators = 100, random _ state =10, subsample=0.8

Table 2 Confusion Matrix

TP	FP
TN	FN

Table 3 The summary performance of the Machine Learning algorithms

Model	Accuracy	Precision	Recall	F1-Score	ROC_AUC
XGBoost	90	0.91	0.90	0.90	90
LGBM	**97**	**97**	**96**	**96**	**97**
Random Forest	75	0.87	0.72	0.72	75
Extra Trees	86	0.89	0.85	0.86	86

images were manually cropped to remove some unwanted background images and noise. Then, they were resized to 128 × 128, flattened, and converted to gray scale before their features were extracted. Firstly, Fig. 2 presents a sample CXR image and their corresponding transformed HOG images for the three different classes (COVID-19, NORMAL, and PTB). The feature vectors were divided into 80:20 train-test split ratio. As discussed in Sect. 5.2, the results of the comparison of the performance of the 4 different machine learning models on the extracted and reduced dataset were presented. The values for all metrics range between 0 and 1. The closer the value of the metrics to 1, the better the model. Table 1 displayed the results of the ML models used on the dataset.

Table 3 presents the performance comparison of the four different classifiers based on precision, recall, F1-Score, and accuracy. The result is based on the test set which is 20% of the dataset. It is observed that LGBM achieved the highest classification report values of 0.91 across all metrics. Its performance value is more consistent than with other models across all metrics. Using F1- score as for comparing all models, Random Forest obtained the least value of 0.72 following closely by Extra Tree with the value of 0.86. LGBM outperformed all other models with a value of 0.97. Random Forest obtained the least recall rate and accuracy values of 0.85 and 86%, respectively. Therefore, making it the least performing model.

6.1 Confusion Matrix

This section further analysis the performances by all models using the confusion matrix for the test set as shown in Fig. 3. It is observed that Fig. 3a representing the confusion matrix for LGBM gave the best recall for each class with minimal error. For XGBoost, out of 106 instances of COVID-19 disease, 97(92%) instances were correctly classified as COVID-19 disease, 3(2%) instances were incorrectly classified as NORMAL, while 6(4%) instances were incorrectly classified as PTB disease. For the class NORMAL, 3(4%) instances were misclassified as COVID-19, while 78 instances (96%) were correctly classified as NORMAL but there was no misclassification with PTB class. Also, for the class PTB, 14(18%) instances were misclassified as COVID-19, while 65 instances (82%) were correctly classified as PTB. Comparing all models on COVID-19 prediction, LGBM performed best with a recall rate of 0.97. Out of 106 instances of COVID-19, 97(92%) were correctly

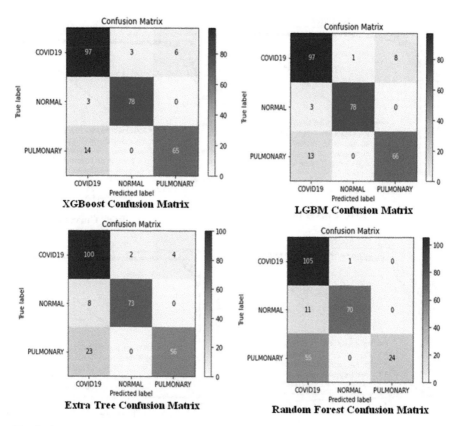

Fig. 3 Confusion Matrix for all models. It is observed the confusion matrix for LGBM gave the best recall for each class with minimal error

classified as COVID-19, 1(1%) instance was misclassified as NORMAL, while 8 (7%) instances were misclassified as PTB. But for the Extra Tree model, only 100 (94%) instances of COVID-19 were correctly classified. 2(2%) instances were misclassified as COVID-19, while 4(4%) instances of COVID-19 were misclassified as PTB. It could be observed that the majority of the misclassification was between COVID-19 and PTB. Hence, our objective has been achieved by building a classification model for COVID-19 disease but separate from PTB.

Analyzing the classification for class NORMAL, there was no misclassification for PTB. All models could distinguish between the images of NORMAL and PTB. The greatest misclassification between NORMAL and COVID-19 was from the XGBoost model where 3(4%) out of 81 instances were misclassified as COVID-19 instead of NORMAL. Analysis of the classification of PTB also shows that there was no misclassification with the class NORMAL image. The greatest misclassification was still between COVID-19 and PTB. For the RF model, 55(70%) out of 79 instances were wrongly classified as COVID-19. LGBM model obtained a recall rate of 0.96 showing a good detection rate.

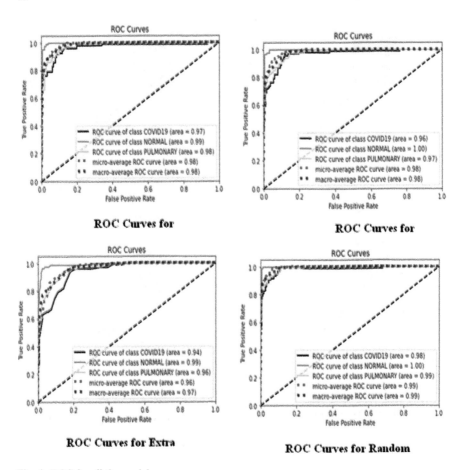

Fig. 4 ROC for all the models

6.2 ROC Curves

The ROC curves for the four models are displayed in Fig. 4 which is the trade-off between true positive and false positive rate having established that it performed best of the models as shown in Fig. 4. The ROC values for all metrics are close to 1 showing a very good classification performance. For example, for LGBM model, the ROC values are 0.96, 1.00, and 0.97 for COVID-19, NORMAL, and PTB, respectively. The overall average is 0.98 showing a good trade-off between recall and precision.

Table 4 shows the result of various models that have compared with the proposed model. The existing model used the same dataset with several conventional Machine Learning and Deep Learning algorithms. Based on the results obtained, the proposed LGBM model performed better using various measurements. The LGBM classifier yield an accuracy of 97%, Precision of 97%, Recall of 96%, F1-score of 96%, and

Table 4 Performance evaluation of the proposed system with the existing state-of-the-art method for COVID-19 pandemic

Models	Method	Accuracy	Precision	Recall	F1-Score	ROC_AUC
[71]	CNNLSTM	92.30			0.93	0.90
[72]	SVM, RF				0.72	0.87
[73]	XGB					0.66
[74]	SVM	95	95	95	94	95
Proposed Model	LGBM	97	97	96	96	97

ROC_AUC of 97%. Therefore, the proposed system using GBLM classifier is considered best method. The dataset size is one of the most limitations associated with the method. The dataset set used was just 520 patients which is very small, this ascribed to the results get using different classifiers. The performance of the proposed model can be enhanced using big dataset with laboratory results from several areas to confirm the outcomes of this work.

7 Conclusion

Cloud-IoT-based is the latest Internet revolution that is a growing field of study, particularly in healthcare. Such remote health care tracking system has developed at such a rate with the rise in the usage of wearable sensors and mobile phones. IoT health surveillance helps to avoid the spread of disease and to allow a correct evaluation of the state of health, even though the doctor is far away. In healthcare, IoT is the main player in supplying patients with better medical services and also facilitates doctors and hospitals. A portable physiological Cloud-IoT-based diagnosis and monitoring system is proposed in this paper, which can continuously screen the body temperature, blood glucose, blood pressure, and other specific room parameters of a patient. The system collects biomedical data from patients through smart technologies and transmits it to the cloud-IoT server to analyze and process the data. Thus, any identification of abnormality in patient information will be reported through the COVID-19 monitoring and alert platform to the patient's physicians. The results of the four used algorithms shown that LGBM performed better with an accuracy of 97%, followed by XGBoost with an accuracy of 90%, and the least inaccuracy of the four algorithms is the Random Forest with 76%. The LGBM model obtained a recall rate of 0.96 in the confusion matrix, which showing a good detection rate. In terms of Recall and Precision, the XGBoost and LGBM performed best with 0.90 and 0.91, respectively, but in F-score, LGBM leads with 0.91. The Cloud-IoT –based healthcare devices can be used for the diagnosis, and obtained data from a person with minor symptoms (Preventive disguises, thermometers, medicines, personalized COVID-19 infection diagnosis, and control kits). Patients were able to submit their general well-being to the IoT-Cloud clinical server database regularly and exchange their relevant data within hospitals, the Center for Disease Control (CDC), national and local healthcare clinics. In future work, more effective security algorithms such as DNA encryption, fully homomorphic encryption, and decryption on the cloud will be implemented to mention a few. It is because data privacy and security are very pressing priorities for IoT-based diagnosis and monitoring of healthcare systems. Mostly because cloud's recorded healthcare data may be subject to various types of security risks.

References

1. Awotunde, J. B., Folorunso, S. O., Jimoh, R. G., Adeniyi, E. A., Abiodun, K. M., & Ajamu, G. J. (2021). Application of Artificial Intelligence for COVID-19 Epidemic: An Exploratory Study, Opportunities, Challenges, and Future Prospects. *Studies in Systems, Decision and Control*, 2021, 358, pp. 47–61.
2. Asai, A., Konno, M., Ozaki, M., Otsuka, C., Vecchione, A., Arai, T., ... Taniguchi, M. (2020). COVID-19 drug discovery using intensive approaches. *International Journal of Molecular Sciences, 21*(8), 2839.
3. Zivkovic, M., Bacanin, N., Venkatachalam, K., Nayyar, A., Djordjevic, A., Strumberger, I., & Al-Turjman, F. (2021). COVID-19 cases prediction by using hybrid machine learning and beetle antennae search approach. *Sustainable Cities and Society, 66*, 102669.
4. Devi, A., & Nayyar, A. (2021). Perspectives on the definition of data visualization: A mapping study and discussion on coronavirus (COVID-19) dataset. *Emerging Technologies for Battling Covid-19: Applications and Innovations*, 223–240. https://doi.org/10.1007/978-030-60039-6_11
5. Joyia, G. J., Liaqat, R. M., Farooq, A., & Rehman, S. (2017). Internet of Medical Things (IoMT): applications, benefits, and future challenges in the healthcare domain. *The Journal of Communication, 12*(4), 240–247.
6. Adeniyi, E. A., Ogundokun, R. O., & Awotunde, J. B. (2021). IoMT-based wearable body sensors network healthcare monitoring system. In *IoT in healthcare and ambient assisted living* (pp. 103–121). Springer.
7. Yang, T., Gentile, M., Shen, C. F., & Cheng, C. M. (2020). Combining point-of-care diagnostics and the internet of medical things (IoMT) to combat the COVID-19 pandemic. *Diagnostics., 10*, 224.
8. Rahman, M. S., Peeri, N. C., Shrestha, N., Zaki, R., Haque, U., & Ab Hamid, S. H. (2020). Defending against the Novel Coronavirus (COVID-19) Outbreak: How Can the Internet of Things (IoT) help to save the World? *Health Policy and Technology, 9*(2), 136–138.
9. Allam, Z., & Jones, D. S. (2020). Pandemic stricken cities on lockdown. Where are our planning and design professionals [now, then, and into the future]? *Land Use Policy, 97*, 104805.
10. Pullano, G., Pinotti, F., Valdano, E., Boëlle, P. Y., Poletto, C., & Colizza, V. (2020). Novel coronavirus (2019-nCoV) early-stage importation risk to Europe, January 2020. *Eurosurveillance, 25*(4), 1.
11. Zhao, S., Lin, Q., Ran, J., Musa, S. S., Yang, G., Wang, W., ... Wang, M. H. (2020). Preliminary estimation of the basic reproduction number of novel coronavirus (2019-nCoV) in China, from 2019 to 2020: A data-driven analysis in the early phase of the outbreak. *International Journal of Infectious Diseases, 92*, 214–217.
12. Christensen, K., Doblhammer, G., Rau, R., & Vaupel, J. W. (2009). Ageing populations: the challenges ahead. *The Lancet, 374*(9696), 1196–1208.

13. Yach, D., Hawkes, C., Gould, C. L., & Hofman, K. J. (2004). The global burden of chronic diseases: overcoming impediments to prevention and control. *JAMA, 291*(21), 2616–2622.
14. Darkins, A., Ryan, P., Kobb, R., Foster, L., Edmonson, E., Wakefield, B., & Lancaster, A. E. (2008). Care Coordination/Home Telehealth: the systematic implementation of health informatics, home telehealth, and disease management to support the care of veteran patients with chronic conditions. *Telemedicine and e-Health, 14*(10), 1118–1126.
15. Awotunde, J. B., Jimoh, R. G., Oladipo, I. D., Abdulraheem, M., Jimoh, T. B., & Ajamu, G. J. (2021). Big Data and Data Analytics for an Enhanced COVID-19 Epidemic Management. *Studies in Systems, Decision and Control*, 2021, 358, pp. 11–29.
16. Chakraborty, C., & Abougreen, A. N. (2021). Intelligent Internet of Things and advanced machine learning techniques for COVID-19. *EAI Endorsed Transactions on Pervasive Health and Technology, 21*, 1–14.
17. Udgata, S. K., & Suryadevara, N. K. (2020). COVID-19: Challenges and advisory. In *The Internet of Things and sensor network for COVID-19* (pp. 1–17). Springer.
18. Muhammad, L. J., Algehyne, E. A., Usman, S. S., Ahmad, A., Chakraborty, C., & Mohammed, I. A. (2021). Supervised machine learning models for prediction of COVID-19 infection using epidemiology dataset. *SN Computer Science, 2*(1), 1–13.
19. Pramanik, P. K. D., Upadhyaya, B. K., Pal, S., & Pal, T. (2019). Internet of things, smart sensors, and pervasive systems: Enabling connected and pervasive healthcare. In *Healthcare data analytics and management* (pp. 1–58). Academic Press.
20. Srivastava, G., Parizi, R. M., & Dehghantanha, A. (2020). The future of blockchain technology in healthcare internet of things security. In *Blockchain cybersecurity, trust and privacy* (pp. 161–184). Springer.
21. Awotunde, J. B., Bhoi, A. K., & Barsocchi, P. (2021). Hybrid Cloud/Fog Environment for Healthcare: An Exploratory Study, Opportunities, Challenges, and Future Prospects. *Intelligent Systems Reference Library*, 2021, 209, pp. 1–20.
22. Darwish, A., Ismail Sayed, G., & Ella Hassanien, A. (2019). The impact of implantable sensors in biomedical technology on the future of healthcare systems. In *Intelligent pervasive computing systems for smarter healthcare* (pp. 67–89). Wiley.
23. Manogaran, G., Chilamkurti, N., & Hsu, C. H. (2018). Emerging trends, issues, and challenges on the Internet of Medical Things and wireless networks. *Personal and Ubiquitous Computing, 22*(5–6), 879–882.
24. Qadri, Y. A., Nauman, A., Zikria, Y. B., Vasilakos, A. V., & Kim, S. W. (2020). The future of Healthcare Internet of Things: A survey of emerging technologies. *IEEE Communications Surveys & Tutorials, 22*(2), 1121–1167.
25. Abikoye, O. C., Bajeh, A. O., Awotunde, J. B., Ameen, A. O., Mojeed, H. A., Abdulraheem, M., Oladipo, I. D., & Salihu, S. A. (2021). Application of Internet of Thing and Cyber-Physical System in Industry 4.0 Smart Manufacturing. *Advances in Science, Technology and Innovation*, pp. 203–217.
26. Awotunde, J. B., Jimoh, R. G., Oladipo, I. D., & Abdulraheem, M. (2021). Prediction of malaria fever using long-short-term memory and big data. *Communications in Computer and Information Science, 2021*(1350), 41–53.
27. Rajabion, L., Shaltooki, A. A., Taghikhah, M., Ghasemi, A., & Badfar, A. (2019). Healthcare big data processing mechanisms: the role of cloud computing. *International Journal of Information Management, 49*, 271–289.
28. Ali, O., Shrestha, A., Soar, J., & Wamba, S. F. (2018). Cloud computing-enabled healthcare opportunities, issues, and applications: A systematic review. *International Journal of Information Management, 43*, 146–158.
29. Rath, M., & Pattanayak, B. (2019). Technological improvement in modern health care applications using Internet of Things (IoT) and proposal of novel health care approach. *International Journal of Human Rights in Healthcare*. https://doi.org/10.1108/IJHRH-01-2018-0007
30. Darwish, A., Hassanien, A. E., Elhoseny, M., Sangaiah, A. K., & Muhammad, K. (2019). The impact of the hybrid platform of internet of things and cloud computing on healthcare systems:

opportunities, challenges, and open problems. *Journal of Ambient Intelligence and Humanized Computing, 10*(10), 4151–4166.
31. Garg, L., Chukwu, E., Nasser, N., Chakraborty, C., & Garg, G. (2020). Anonymity preserving IoT-based COVID-19 and other infectious disease contact tracing model. *IEEE Access, 8*, 159402–159414.
32. Din, S., & Paul, A. (2020). Erratum to "Smart health monitoring and management system: Toward autonomous wearable sensing for Internet of Things using big data analytics [Future Gener. Comput. Syst. 91 (2019) 611–619]". *Future Generation Computer Systems, 108*, 1350–1359.
33. Otoom, M., Alshraideh, H., Almasaeid, H. M., López-de-Ipiña, D., & Bravo, J. (2015). Real-time statistical modeling of blood sugar. *Journal of Medical Systems, 39*(10), 123.
34. Alshraideh, H., Otoom, M., Al-Araida, A., Bawaneh, H., & Bravo, J. (2015). A web based cardiovascular disease detection system. *Journal of Medical Systems, 39*(10), 122.
35. Wu, F., Wu, T., & Yuce, M. R. (2019). An internet-of-things (IoT) network system for connected safety and health monitoring applications. *Sensors, 19*(1), 21.
36. Gope, P., & Hwang, T. (2015). BSN-Care: A secure IoT-based modern healthcare system using body sensor network. *IEEE Sensors Journal, 16*(5), 1368–1376.
37. Verma, P., & Sood, S. K. (2018). Cloud-centric IoT based disease diagnosis healthcare framework. *Journal of Parallel and Distributed Computing, 116*, 27–38.
38. Xu, B., Da Xu, L., Cai, H., Xie, C., Hu, J., & Bu, F. (2014). Ubiquitous data accessing method in IoT-based information system for emergency medical services. *IEEE Transactions on Industrial Informatics, 10*(2), 1578–1586.
39. Banaee, H., Ahmed, M. U., & Loutfi, A. (2013). Data mining for wearable sensors in health monitoring systems: a review of recent trends and challenges. *Sensors, 13*(12), 17472–17500.
40. Folorunso, S. O., Awotunde, J. B., Ayo, F. E., & Abdullah, K. K. A. (2021). RADIoT: The Unifying Framework for IoT, Radiomics and Deep Learning Modeling. *Intelligent Systems Reference Library*, 2021, 209, pp. 109–128.
41. Zhang, M. W., Ward, J., Ying, J. J., Pan, F., & Ho, R. C. (2016). The alcohol tracker application: an initial evaluation of user preferences. *BMJ Innovations, 2*(1), 8–13.
42. Zhang, M. W., & Ho, R. (2017). Smartphone application for multi-phasic interventional trials in psychiatry: Technical design of a smart server. *Technology and Health Care, 25*(2), 373–375.
43. Mahapatra, B., Krishnamurthi, R., & Nayyar, A. (2019). Healthcare models and algorithms for privacy and security in healthcare records. In *Security and Privacy of Electronic Healthcare Records: Concepts, Paradigms and Solutions* (p. 183). IET.
44. Rathee, D., Ahuja, K., & Nayyar, A. (2019). Sustainable future IoT services with touch-enabled handheld devices. *Security and Privacy of Electronic Healthcare Records: Concepts, Paradigms and Solutions*, 131.
45. Nguyen, T. T. (2020). Artificial intelligence in the battle against coronavirus (COVID-19): A survey and future research directions. Preprint. https://doi.org/10.13140/RG.2.2.36491.23846/1.
46. Nayyar, A., Gadhavi, L., & Zaman, N. (2021). *Machine learning in healthcare: review, opportunities and challenges*. Machine Learning and the Internet of Medical Things in Healthcare, 23–45.
47. Rao, A. S. S., & Vazquez, J. A. (2020). Identification of COVID-19 can be quicker through artificial intelligence framework using a mobile phone–based survey when cities and towns are under quarantine. *Infection Control & Hospital Epidemiology, 41*(7), 826–830.
48. Allam, Z., & Jones, D. S. (2020, March). On the coronavirus (COVID-19) outbreak and the smart city network: universal data sharing standards coupled with artificial intelligence (AI) to benefit urban health monitoring and management. In Healthcare (Vol. 8, No. 1, p. 46). : Multidisciplinary Digital Publishing Institute
49. Fatima, S. A., Hussain, N., Balouch, A., Rustam, I., Saleem, M., & Asif, M. (2020). IoT enabled smart monitoring of coronavirus empowered with fuzzy inference system. *International Journal of Advance Research, Ideas and Innovations in Technology, 6*(1), 188–194.

50. Peeri, N. C., Shrestha, N., Rahman, M. S., Zaki, R., Tan, Z., Bibi, S., ... Haque, U. (2020). The SARS, MERS and novel coronavirus (COVID-19) epidemics, the newest and biggest global health threats: what lessons have we learned? *International Journal of Epidemiology, 49*(3), 717–726.
51. Sharma, K., Singh, H., Sharma, D. K., Kumar, A., Nayyar, A., & Krishnamurthi, R. (2021). Dynamic models and control techniques for drone delivery of medications and other healthcare items in COVID-19 hotspots. *Emerging Technologies for Battling Covid-19, 324*, 1.
52. Ayo, F. E., Awotunde, J. B., Ogundokun, R. O., Folorunso, S. O., & Adekunle, A. O. (2020). A decision support system for multi-target disease diagnosis: A bioinformatics approach. *Heliyon, 6*(3), e03657.
53. Ayo, F. E., Ogundokun, R. O., Awotunde, J. B., Adebiyi, M. O., & Adeniyi, A. E. (2020, July). *Severe acne skin disease: a fuzzy-based method for diagnosis*. Lecture Notes in Computer Science (including subseries Lecture Notes in Artificial Intelligence and Lecture Notes in Bioinformatics), 2020, 12254 LNCS (pp. 320–334).
54. Oladele, T. O., Ogundokun, R. O., Awotunde, J. B., Adebiyi, M. O., & Adeniyi, J. K. (2020, July). *Diagmal: A Malaria coactive neuro-fuzzy expert system*. Lecture Notes in Computer Science (including subseries Lecture Notes in Artificial Intelligence and Lecture Notes in Bioinformatics), 2020, 12254 LNCS (pp. 428–441).
55. Awotunde, J. B., Folorunso, S. O., Bhoi, A. K., Adebayo, P. O., & Ijaz, M. F. (2021). Disease Diagnosis System for IoT-Based Wearable Body Sensors with Machine Learning Algorithm. *Intelligent Systems Reference Library, 2021, 209*, pp. 201–222.
56. Ajagbe, S. A., Idowu, I. R., Oladosu, J. B., & Adesina, A. O. (2020). Accuracy of machine learning models for mortality rate prediction in a crime dataset. *International Journal of Information Processing and Communication (IJIPC), 10*(1&2), 150–160.
57. Pramanik, P. K. D., Nayyar, A., & Pareek, G. (2019). WBAN: Driving e-healthcare beyond telemedicine to remote health monitoring: Architecture and protocols. In *Telemedicine technologies* (pp. 89–119). Academic Press.
58. Li, J., Chen, Z., Nie, Y., Ma, Y., Guo, Q., & Dai, X. (2020). Identification of symptoms prognostic of COVID-19 severity: multivariate data analysis of a case series in Henan province. *Journal of Medical Internet Research, 22*(6), e19636.
59. Otoom, M., Otoum, N., Alzubaidi, M. A., Etoom, Y., & Banihani, R. (2020). An IoT-based framework for early identification and monitoring of COVID-19 cases. *Biomedical Signal Processing and Control, 62*(102149).
60. Rabby, M. K. M., Alam, M. S., & Shawkat, M. S. A. (2019). A priority-based energy harvesting scheme for charging embedded sensor nodes in wireless body area networks. *PLoS One, 14*(4), e0214716.
61. Chen, C. M., Xiang, B., Wu, T. Y., & Wang, K. H. (2018). An anonymous mutual authenticated key agreement scheme for wearable sensors in wireless body area networks. *Applied Sciences, 8*(7), 1074.
62. Devi, V. A., & Nayyar, A. (2021). Evaluation of geotagging Twitter data using sentiment analysis during COVID-19. In *Proceedings of the Second International Conference on Information Management and Machine Intelligence* (pp. 601–608). Springer.
63. Kumar, A., Krishnamurthi, R., Nayyar, A., Sharma, K., Grover, V., & Hossain, E. (2020). A novel smart healthcare design, simulation, and implementation using healthcare 4.0 processes. *IEEE Access, 8*, 118433–118471.
64. Medina, J., Espinilla, M., García-Fernández, Á. L., & Martínez, L. (2018). Intelligent multi-dose medication controller for fever: From wearable devices to remote dispensers. *Computers & Electrical Engineering, 65*, 400–412.
65. Umayahara, Y., Soh, Z., Sekikawa, K., Kawae, T., Otsuka, A., & Tsuji, T. (2018). A mobile cough strength evaluation device using cough sounds. *Sensors, 18*(11), 3810.
66. Ichwana, D., Ikhlas, R. Z., & Ekariani, S. (2018, October). Heart rate monitoring system during physical exercise for fatigue warning using non-invasive wearable sensor. In *2018 International Conference on Information Technology Systems and Innovation (ICITSI)* (pp. 497–502). IEEE.

67. Askarian, B., Yoo, S. C., & Chong, J. W. (2019). Novel image processing method for detecting strep throat (streptococcal pharyngitis) using smartphone. *Sensors, 19*(15), 3307.
68. Gaidhani, A., Moon, K. S., Ozturk, Y., Lee, S. Q., & Youm, W. (2017). Extraction and analysis of respiratory motion using wearable inertial sensor system during trunk motion. *Sensors, 17*(12), 2932.
69. Krishnamurthi, R., Gopinathan, D., & Nayyar, A. (2021). A comprehensive overview of fog data processing and analytics for healthcare 4.0. Fog computing for healthcare 4.0 environments (pp. 103–129).
70. Pedregosa, F., Varoquaux, G., Gramfort, A., Michel, V., Thirion, B., Grisel, O., . . . Duchesnay, E. (2011). Scikit-learn: Machine learning in Python. *The Journal of machine Learning research, 12*, 2825–2830.
71. Mohammed, M. A., Abdulkareem, K. H., Al-Waisy, A. S., Mostafa, S. A., Al-Fahdawi, S., Dinar, A. M., . . . Díez, T. (2020). Benchmarking methodology for selection of optimal COVID-19 diagnostic model based on entropy and TOPSIS methods. *IEEE Access, 8*, 99115–99131.
72. de Moraes Batista, A. F., Miraglia, J. L., Donato, T. H. R., & Chiavegatto Filho, A. D. P. (2020). *COVID-19 diagnosis prediction in emergency care patients: a machine learning approach.* medRxiv.
73. Schwab, P., Schütte, A. D., Dietz, B., & Bauer, S. (2020). *Predcovid-19: a systematic study of clinical predictive models for coronavirus disease 2019.* arXiv preprint arXiv:2005.08302.
74. Abdulkareem, K. H., Mohammed, M. A., Salim, A., Arif, M., Geman, O., Gupta, D., & Khanna, A. (2021). Realizing an effective COVID-19 diagnosis system based on machine learning and IoT in smart hospital environment. *IEEE Internet of Things Journal.* https://doi.org/10.1109/JIOT.2021.3050775

AIIoT for Development of Test Standards for Agricultural Technology

Puneet Kumar Aggarwal, Parita Jain, Poorvi Chaudhary, Riya Garg, Kshirja Makar, and Jaya Mehta

1 Introduction

The extensive popularity and use of the Internet in society have been witnessed for many years. This has prefaced immeasurable gains for firms and residents across the globe. The influential interest of this reform has created the potential to bridge the client and producer gap in actual time. In today's world, the Internet of Things (IoT) is assuring identical gain with the establishment of revolutionary technologies and providing a method to magnify the person's knowledge and ability by improving the functioning of the traditional mechanisms and practices. IoT can help in finding solutions in the fields of healthcare, retail, security, agriculture, and a lot more. The introduction of IoT in farming and agricultural activities is of great use as continuous monitoring and observation are required in this feel to expect a better yield, a healthy product, and reduction in wastage.

The system of agricultural IoT includes precision farming, livestock, and greenhouses are some of the features which are grouped into different domain names. These activities are being monitored with the help of IoT-based gadgets like sensors and robots using the technique of wireless or wired networks that may help the farmers to acquire records of the weather conditions, humidity, soil texture, air tracking, temperature tracking, water monitoring, disease monitoring, place monitoring, environmental situations tracking, pest tracking, and fertilization monitoring and crop growth. The most thrilling part of the evolved system is the clam which

---

P. K. Aggarwal (✉)
ABES Engineering College, Ghaziabad, India

P. Jain
KIET Group of Institutions, Ghaziabad, India

P. Chaudhary · R. Garg · K. Makar · J. Mehta
HMRITM, Delhi, India

© The Author(s), under exclusive license to Springer Nature Switzerland AG 2021
F. Al-Turjman et al. (eds.), *Intelligence of Things: AI-IoT Based Critical-Applications and Innovations*,
https://doi.org/10.1007/978-3-030-82800-4_4

offers a reliable and organized observation through the established networks along with the integration of artificial intelligence and cloud computing. IoT additionally monitors the situations of the environment to assure the readings of air pollution, water pollution, temperature, and detrimental radiations. On the other hand, artificial intelligence is one of the fastest emerging fields of technology which has the potential to change modern agriculture.

The artificial intelligence (AI) based tool and algorithms which when fed with data can predict the amount of usage of fertilizers, water, heat, and other factors for the better yield of crop and better product quality according to the test standards. The reason for this chapter is to exhibit and know how settling on choices with cutting-edge information-based agribusiness can increase the accessibility, health benefits, and profit margins.

The main motive is to automate the agriculture-based business and the test structures so that the end product is more cost efficient, requires less human interaction and appropriate usage of resources. Various surveys have shown that there is a huge requirement for a platform which can intake the data that is produced on a farm on a daily basis as the amount of data generated seems to increase by each passing day. The data that yields the information can be transformed into productive choices just by the help of AI and IoT. Not just in the growth and development of the firms but also to provide healthy and safe food is the real motive of the agricultural businesses to open their doors for AIIoT.

AIIoT is nothing but the combination of the two broad fields which are AI and IoT. Further in this chapter we are going to discuss the importance of AIIoT in the agricultural test standards and how it is important for sustainable agriculture. As we all know farmers produce food and if the production process which includes cultivation, irrigation, and packaging is done in a disciplined order, then the product will be both fit for consumption and cost efficient.

The discipline can be achieved only by following the guidelines laid by international standards for specific tools, machines, and production techniques. These standards need to meet the required marks to prove whether they are fit for consumption or not. This makes us follow the ISO standards which are made to certify the products which are grown in farms and to verify that all the processes which the crop undergoes in the farm are according to the laid guidelines by the international organization.

The rest of the chapter is organized into different sections. Sect. 1.1 discusses about the concept of AIIoT and how it works. In Sect. 2 how IoT uses AI is discussed. Then, various applications of IoT and AI are discussed and described in Sect. 3. Sect. 4 discusses about the how testing is formed in agriculture. The various testing standards followed in agriculture are discussed in Sect. 5 of the chapter. The main ISO standard that is followed for testing various software used in agriculture is described in Sect. 6. Then in Sect. 7 conclusions are drawn from the chapter. Finally references at last.

1.1 Artificial Intelligence-Internet of Things (AIIoT)

Today with the advancement of technology, the new era of the Internet has made the lives of people quite easy. As a result, now we can use the Internet and network systems everywhere starting from homes, schools to offices, and companies. However, this simple use of technology is not enough. People now have started to connect their all basic day-to-day life things with the Internet to reduce their time and make life easier and effective at the same time. So, this activity of connecting things with the Internet was introduced with the concept of Internet of Things. It is estimated that approximately more than 15 billion things are connected to the Internet. Despite such a large number of things connected it makes only 1% of all the things available that could potentially be connected with the Internet.

With the increase of dependency on the internet, various internet threats are refereed as disadvantages to IoT such as threat of security. Minimizing the threats such that it will not affect the efficiency of the smart devices, the combination of AI and IoT is introduced. AI enables the systems to perform reasoning like human beings and hence make "smart decisions". With the spread of AI in applications like online shopping, email services, etc. make it used by every person somewhere in his daily life. AIIoT as the name signifies is the combination of two major technologies AI and IoT. While IoT represents the connection of various systems on the Internet, AI is responsible for the decision-making process which means it acts like the brain of the system. The combination of two forms of fully connected systems functions like self-correcting, self-analysing, etc. [1, 2].

This mixture supports the working of different complex algorithms like machine learning as well as various analytical techniques such as big data analytics which makes the work easier and at the same time efficient. This combination of Internet of things and artificial intelligence makes the life "smart". All the functions that occur automatically such as the smart reply suggestions on Gmail personalized suggested searches on LinkedIn and other websites is all made achievable with the help of AI and IoT. This concept also involves the idea of Internet of Everything which refers to connections of everything with everyone.

This idea basically aims to connect things virtually as well as physically whether it is living or non-living. As we have seen the virtual connection is made possible due to Internet of Things and similarly to make the things connect physically there comes the concept of CPS (Cyber Physical System). In order to make a system AIIOT enabled, the first and most important step is to introduce artificial intelligent in IoT based systems. Due to this today the systems have become more economic, intelligent, and smarter at the same time. The major aspect that needs to kept in mind while building IoT systems is its architecture and its flexibility to work in given conditions. The major thing that drives these processes is the availability and exchange of data. More is the amount of data more is needed for automated technologies in order to make processes faster to interpret. While on the one hand AIIoT helps to process the data, on the other hand, it makes it consumes a lot of power and constant maintenance. Through these points seem to us the disadvantages but these are manageable at the same time and looking at the results that are produced all the input and investment seems worth it.

1.1.1 Internet of Things (IoT)

As the world is moving towards technology IoT is a subject that cannot be ignored. IoT can be referred to as a system that connects various digital and computer devices and makes human–computer interaction much easier. It is a term that creates a notion that all the IoT devices have the capability to inspect and collect the information from the real world and pass it on to Internet that is virtual world where it can be used positively for other applications and interesting projects. The concept of "smart devices" or "smart home" has come into existence because of IoT. All the appliances that we use today or will be using in future (such as lighting fixtures, thermostats, home security systems and cameras etc.) that support one or more common ecosystems or environments use IoT in some direct or indirect way. Today each and every common person in the world is familiar with the word IoT because it is making our life much easier and comfortable [3].

It consists of a well-connected system made up of smart devices such as sensors, communication software and hardware, and some processors to act appropriately on the data received from their ecosystems. These smart devices share the data among them that they have collected by connecting to a main device also called "IoT gateway" where the data collected is further sent to be analysed. These devices are related only by the means of information or data that is shared between them. The speciality of these smart devices is that they are designed in such a way that they not require manual—aid to perform their functions, though humans can intervene to assure their proper working such as providing the instructions, managing the settings, accessing the data, etc. They use various connectivity, networking, and communication protocols to successfully form the applications that we use today. The most important thing about these devices is that these devices provide the ideal service as system not as an independent individual that is they cannot work individually; they need a connected system and network in order to perform function [4, 5].

IoT is especially important for large scale businesses where all the work cannot be handled manually. Also, it costs much less than the human labour and is more efficient. Hence it reduces time, cost, and effort. Apart from setting large scale businesses it is used to manage already set up businesses as it helps to look into the smallest details for example; how the systems actually work, whether all the things are working as expected, the growth rate, the problems in the system, logistic operations, etc. It also helps in effective decision making and generates more revenue [6].

Architecture of IoT

IoT is not only the technology that consists of devices that connect humans and Internet. But it is actually the technology that enables the building of complete systems that are capable of sensing information from the human world and performing all the functions without human intervention. Actually, this system is quite complicated to be implemented if it is not well built [7, 8]. Therefore, the architecture of IoT plays an important role in simplifying the implementation part of the system. This also reduces the number of sources invested in the process. Understanding the architecture makes us understand what the concept of IoT actually means. IoT architecture is the system that contains number of elements: sensors, protocols, actuators, cloud services, and layers.

The three IoT architecture layers are as follows:

1. IoT Device Layer
2. IoT Gateway Layer
3. IoT Platform Layer

Considering these layers of IoT is very important as they are a very crucial part of the architecture. Besides these layers IoT architecture includes functionality, scalability, availability, and maintainability. The four main stages of IoT architecture are described in Fig. 1 as shown below:

1. *Stage 1: Sensors and Actuators:*
   The ability of sensors is to recognize and convert the information available in the real world for analysis. This stage is important to start by actually collecting the information. Actuators are even more advanced and better than sensors as they are able to intervene into the reality and adjust the information accordingly. They generally control the physical environment. For example, they are capable to switch off the lights and fans, etc.
2. *Stage 2: Internet Gateway and Data Acquisition Systems:*
   The data in the sensors now need to be sorted, aggregated, and converted into digital signals. Data acquisition systems are required in order to perform these aggregations and conversions. DAS work by connecting to sensors while Internet gateways connect to Wi-Fi and wired LANs for further processing.

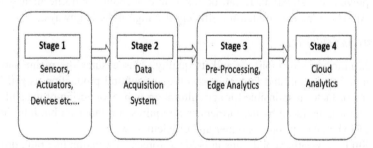

Fig. 1 Four Stages of IoT Architecture

3. *Stage 3. Edge IT:*

Now the data is prepared to send into the computer world. Edge IT here play a vital role in pre-processing and analysing the data one last them before it is transferred into IT world. This stage connects IOT to the IT world and establishes an indirect connection between human and computer world too.

4. *Stage 4. Data Centre and Cloud:*

This is actually the management stage that helps in storing and managing the already analysed data. The main task is visualization and management. Here the data is further sent to cloud-based systems. After this stage the process again repeats itself.

Applications of IoT

It can be used in almost all the fields like educational system, railways, airports, bus stand for display notifications and many more. Some of the applications are described as [9–11]:

1. *Smart Cities:*

   (a) Monitoring of parking areas availability in the city. The maturing of IoT has profoundly affected the commencement of the technology in terms of smart parking. From parking splotches to real-time parking maps, the infusion of IoT has availed everything further accelerated and effortless. Travelling becomes easier when solace is intricate. Incorporation of IoT platforms has upgraded the level of magnificence.

   (b) Monitoring of vehicles and pedestrians to optimize driving and walking routes. Safety is predominant especially when it is just a click away. IoT platform has sanctioned trackers or applications which can allow users to supervise the circumstances on roads. Real-time monitoring of driving can reduce the life risks exceedingly.

   (c) Intelligent Highways that warns according to climate conditions and unexpected events like accidents or traffic jams. The census of accidents reveals that the supreme number of accidents take place on highways. With the assimilation of IoT platforms diverse sensors, equipment can be used to prevent the situation. It can be taken into account for traffic management. Alerts can be convoluted for efficient driving on the Highways.

2. *Smart Agriculture:*

   (a) Monitoring soil moisture and its quality. This ensures the maintenance of nutrients in soil and thus increases the agricultural production as soil is the major factor responsible for agricultural growth. Monitoring of the soil is the major issue faced by the farmers as it requires constant care but introduction of IoT makes it much simpler and efficient as well.

   (b) Study of weather conditions in fields to forecast formation like rain, drought, snow, or wind changes. Climate plays a major role in farming and growth of crops. Inappropriate knowledge of weather forecast can lead to deterioration of both quality and quantity of crops. IoT allows real-time monitoring of weather conditions with help of the sensors placed in the farms. Sometimes these sensors also inspect the condition of the crops and the environmental conditions thus making it easy for the farmers to decide which crops to grow in which season.

(c) It also helps in control of humidity and temperature levels in crops to prevent fungus and other microbial contaminants. It brings more precision or accuracy to the farming. Monitoring of different factors in farms like vehicles, tools utilized, weather conditions, soil nutrients, and other important factors lead to increment in the accuracy in farming which in turn assures good quality as well as quality of crops. This also allows the farmers to make quick and correct decisions. This also ensures profit in terms of the investment made.

3. *Smart Homes:*

(a) Automatic Switching on and off appliances like fans, lights, etc. is done to save energy and to reduce the risk of hazards or accidents. For example, it detects the water or gas leakage within the house with the help of sensors and then the IoT ecosystem is altered and automatically the home appliance is turned off. All the process of detecting as well as taking the required action is completed by IoT system on its own which makes the home even smarter.

(b) Detection of windows and doors openings to prevent intruders is done which increases the security of the home and its members with help of security cameras, and other sensors which notify the user if anything goes wrong or unexpected. Other than this it can also detect smoke. The smoke detectors allow the sensors to create a wireless network and connect the system to an app so that the user has all the updates of what is going on even when he is not in the house. Thus, if the smart home does not have any caretaker but then an IoT based smart home can take care of itself.

4. *Medical Fields:*

(a) Monitoring of conditions of patients inside hospitals can be done easily. Healthcare sector has unleashed the ultimate remedy for patients' safety and health with their collaboration with IoT. It has become further straightforward to store the data of patients more dynamically rather than maintaining columns on pages. With the familiarity of these devices patients are attended with more convenience and peace.

(b) Also, monitoring and controlling of conditions inside freezers for storing medicines and vaccines can be done easily. Pharmacy availability on a click has reformulated the circumstances to a further extent of proficiency. From applications of medicines providers to the results of tests, the evolution has been brisk and successful. Certain requisitions for hospital equipment, pumps, machines, etc. are now available. These devices keep a track of hospital requirements so that every patient should get the finest treatment.

1.1.2 Artificial Intelligence (AI)

Artificial intelligence actually refers to mainly making the systems able to make smart decisions such as a human brain in such a way that the system can execute most of its tasks automatically. This makes the interpretation of the data easier and enhances the process of decision making. This means there is no need for constant manual surveillance as it makes the machines more automatic and performs self-capable tasks. Apparently, they cannot make the decisions as creative as humans but then why do we need them? It is because of the increase in the amount of data prevalent in the industry.

As the flow of data is increasing there is less time available to interpret it and make the decisions and comparisons within limited time. So, in order to make the processes fast artificial intelligence comes into play. Hence it reduces human work as well as effort and makes life easier. It is currently being used in all the domains like computer science, business, marketing, statistics, etc. Not only the data available is in huge amounts but also it is quite unstructured. Thus, the manual sporting of the data requires a lot of time and effort while if this sorting is done using some algorithms in the machines it will make it simpler and more efficient.AI mainly depends on techniques and algorithms generally built using machine learning by the developers. Many methods as well tools have now been developed which makes the process of decision making more accurate and predictable. Generally, people think AI refers to self-driving cars or robots but it is used in more basic or root levels such as connecting the systems and devices to the Internet and with each other, sorting the data and making suitable comparisons and thus provide user-oriented results.

1.1.3 Machine Learning (ML)

Machine learning can be described as a subcategory of the Artificial Intelligence (AI) field which has its main focus on examining and recognizing patterns and arrangements in data to facilitate features such as training, thinking, decision making, learning, and researching without interference. Machine learning allows the user to pack an enormous sum of data with a computer algorithm, allows the computer to examine and analyse the data to make recommendations based on the input received. If some features require redesigning, they are classified and improved for a better design for the future [12, 13].

The main aim of the technology is to produce an easy to use a mechanism which works on making decisions by the computational algorithm. Variables, algorithms, and innovations are accountable for making decisions. Awareness towards the solution is needed for the better learning of the systems working and thus helps in understanding the path to reach the result.

In the initial stages of implementation of the algorithm, input or data is provided to the machine provided that the result is already known for that set of feed data. The changes are then made to produce the result [14–16, 37]. The efficiency of the result

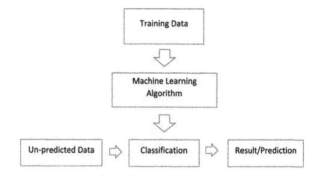

Fig. 2 Machine Learning Algorithm Process

depends on the amount and the quantity of the data that is fed to the machine as shown in Fig. 2.

2 Internet of Things Using Artificial Intelligence

The systems using only Internet of Things are the systems which are quite complex as it uses a lot of mathematical calculations and a number of algorithms and thus take longer to make accurate decisions and sometimes cause numerous threats to security of users. The usage of artificial intelligence in these systems can solve the above problems. It makes the system more reliable and reduces the time as well. An AIIoT system gives more specific comments according to the user. The major change that occurs is in the architecture of the system as it supervises and ensures the performance of the system is shown in Fig. 3. But at the same time, it should be made sure that there is no change in the original data received from the user.

Therefore, AI methods cannot be placed at the same level in IoT architecture. Another advantage of using AI in IoT enabled systems is that it makes machines run able in almost all the servers as AI methods are suitable to most of the servers. The major advantage of such systems is that the information can be reused. Like in the human brain once an information is stored in it can be used at any point of time similarly in AIIoT systems given both AI and IoT the information once entered in the system is stored in server rooms which can be referred and reused at any point in future. This helps to look at things with a broader perspective [17–19].

There is a proper process with which the IoT systems can get enabled with artificial intelligence work. In order to understand the working of these systems the following three components must be understood:

1. Preliminary Communication: the communication that takes place when there is transfer of data from the human world to the system. In this stage there is the collection of data from the outside world. The data sent is initially pre-processed for checking whether it is corrupt or not. This transmission of data takes place over the Internet and the devices connected to the web. Once the data is collected

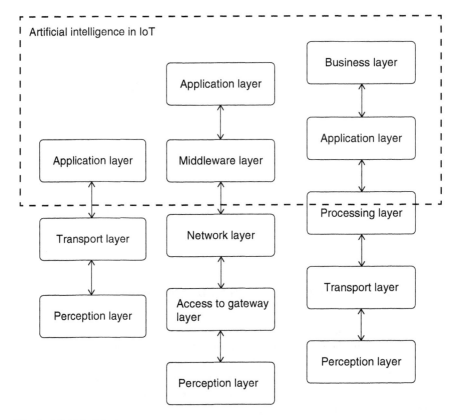

Fig. 3 AIIoT Architecture

by the IoT systems then it becomes ready for the application of artificial intelligence methods.
2. Context Communication: the communication that takes place when the received data is interpreted and then the AIIoT system is made to make decisions accordingly. Here the main purpose is analysis of the information received and a proper decision making by AI developed systems and then sending responses to IoT machines. Generally, the response produced by AI systems includes digits which are not properly understood by the IoT systems without a context.
3. Internal Communication: the communication that takes place when the response generated by the system is well understood by the interconnected systems by the server.

2.1 Advantages of AIIoT

There are various advantages of enabling Internet of Things using artificial intelligence which can be described as follows:

1. Increase in Efficiency: AI when enabled with IoT makes the systems more reliable and accurate as it collects the data and figures the patterns in it and draws suitable conclusions and comparisons. The AI and ML methods perform suitable operations such that the results generated are ideal. This is usually done by predicting the possible conditions that can lead to failure. While IoT predicts the redundancy in the data and reduces the time.
2. Improves the Risk Management: As the AIIOT systems work on the principle of prediction hence it predicts the risks or probability of failure within the system which makes it easier to remove the risks and provide security. For example, these systems help to manage financial risks or cyber threats.
3. Improves the scalability and flexibility: This means Ai present in the Iot devices help to analyse the huge amount of data easily and makes the flow of information from one device to another easier. IOT devices vary from huge computers to small micro-sensors. So to make the flow of data easier from large to small devices and vice versa AIIOT methods are most suitable.

2.2 Applications of AIIoT

Some of the applications of AIIoT can be defined as:

1. Automatic alarms: its functionality is divided into three components where the mobile phones act as the smart objects which actually collect the data from the outside world using the sensors that are built within the devices, there is a server which plays the role of IoT system and then the micro-services which involve the functions of artificial intelligence. Each and every smart device has the presence of an identifier that connects with the central server and pulls and pushes its requests.
2. Robotic Vacuum Cleaners: The basic function of this automatic vacuum cleaner is to remember the layout or architecture of a particular room or part which it has to clean, and then move on its own accordingly. Here the layout of the room acts as the vital information from the real world that must be provided to the IoT System.
3. Self-driving cars by Tesla: It involves the application of artificial intelligence which makes cars think like human drivers and replace it with a machine.

3 Role of AI and IoT in Agriculture

With the blooming years there have been tremendous developments in the field of agriculture. Agriculture has the power to shake the economy of a country in an increasing or a decreasing state. Agriculture is directly proportional to the equipment and the efforts that the farmers put into. Technologies like IoT and AI have made it possible to revolutionaries this field as shown in Fig. 4.

From the above figure it is very clear that many new and great changes are adapted by the agriculture field. If IoT is used to analyse the data piled up, then AI is the decision maker for the next step. AI contributed in prediction making on the basis of the data collected. With the integration of AI and IoT, the process of agriculture will be more efficient and reliable and can target a large number of new ideas for higher production of the crops [20, 21]. These techniques can be time efficient as well as they can fulfil modern demands of the farmer and people. Purity is the major concern of people nowadays which is completely ensured by the modern techniques accepted by the agriculture industry.

3.1 Role of IoT in Agriculture

The Internet of Things is the supreme and salient feature in the field of agriculture. Advanced machinery is also engaged in agriculture which requires tons of data to be associated and modified to communicate with this data the Internet of Things plays a sophisticated role. IoT has the potential to involve the data over the network without engaging the human's involvement. In agriculture water, climate and sunlight plays a major role for effectively increasing the production of crops. These factors are not stable but changed according to the conditions to tackle this problem [22, 23].

It is important to have a command over these situations which is possible only through data collected and desired outputs on the same. With the help of the Internet of Things access can be provided to farmers to have complete and necessary information about crops, production, necessary pesticides, insecticides, etc. In recent

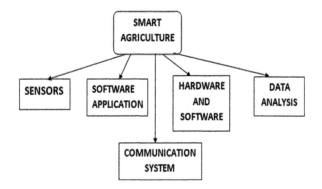

Fig. 4 Smart Agriculture

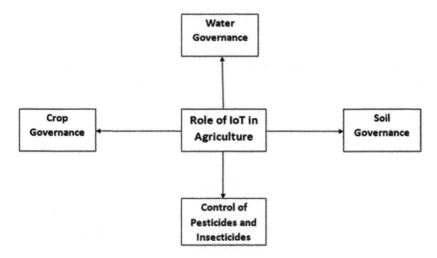

Fig. 5 Role of IoT in Agriculture

years, immense improvement in the field of agriculture has been witnessed and with the Internet of Things there is far more way to go in terms of advancement. The role of IoT can be described as shown in Fig. 5.

3.1.1 Applications of IoT in Agriculture

There are several applications related to the role of the Internet of Things in the field of agriculture which are being practiced in many areas of the world.

1. Weather Monitoring: Every crop has its different climatic conditions which should be fulfilled for producing quality crops and increases the production. The Internet of Things allows the farmers to have knowledge about the weather at all the possible times by having various sensors which will work by collecting data from the environment and monitor that data to provide necessary information to the farmer to select the desired crop for producing and yield the crops. These sensors are capable of handling real-time data and collect information for rainfall, humidity, air quality, etc.
2. Livestock Supervision: IoT enabled devices will be responsible for monitoring the effective presence of livestock in the field. This monitoring includes supervision of livestock's health, total count of livestock, their proper vaccinations so as to keep them healthy.
3. Field Observation: The quality of the field where the crops are grown is a very important factor to take care of. Soil quality according to the desired crop, properly maintained field for the next production, etc. are some of the factors which are monitored by the IoT enabled devices.

These sensors provide sensitive data to the farmers regarding the field which should be improved by the farmers for increasing the yield of the crops and produce good quality products from this field. The precautions which can be adopted with the help of the Internet of Things can result in increasing the production cost, reducing losses, etc.

3.2 Role of AI in Agriculture

Till now we have discussed the role of IoT in agriculture where we came to know that various factors like climate, productivity, etc. can be easily determined with the Internet of Things. Artificial Intelligence has proved to be another important factor in the field of agriculture. To predict the desired environment for better production, the data collected must be analysed and manipulated properly to conduct a predictive approach. Population increase is a permanent factor which will always go on increasing and to feed the whole population the agriculture sector needs to increase its output in terms of crops and production [24, 25].

Engaging Artificial Intelligence in agriculture will help in integrating new and advanced technologies and to provide regular information to the farmers about the crops, its defects, and ways to improve the production. Artificial Intelligence has the capacity to predict various barriers of farming to the farmers which can accelerate the production of the crops as shown in Fig. 6.

3.2.1 Applications of AI in Agriculture

There are only few applications that are capable to fall under this section, being discussed as follows:

1. Agriculture Robots: It is not a very common application but is practiced in few areas of the world. These robots are also known as agribots. Their major role is

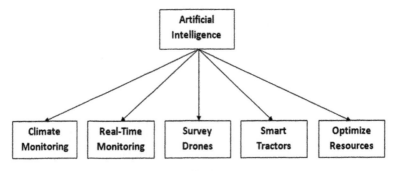

Fig. 6 Role of AI in Agriculture

fulfilled when manual interventions are no more available and they provide efficiency in the crop production.
2. Crop and Soil Supervision: Artificial Intelligence provides prediction of the quality of crop and what level of performance will that crop manage to pull off with that quality before reaping of crops. Artificial Intelligence allows the farmers to have a proper survey about the quality of the soil so that the decision can be made on the type of crop to yield and if there is some factor which is affecting the soil quality, it can be improved timely before it causes more harm to other fields. Many organizations are investing in real-time systems that can work on these applications; robot drones are the best example of the same. These systems will have the capability to provide a quick overview of the field and necessary treatments for the same.
3. Predictive Analysis: There are several barriers in agriculture and no farmer will be willing to harm customers with bad quality crops or products for the same predictive analysis can come into picture to predict several issues and solutions to the same. In this particular application machine learning plays a major role to be integrated to have a keen supervision on the agricultural field and environment around it. These devices are made to track each and every step-in agriculture and the data collected from the same will be used for prediction on the production crop.

3.3 Future Aspects of IoT and AI in Agriculture

The future scope of IoT and AI in agriculture includes various applications in which it can be applied [26].

1. With the brisk speed of technology these are new theories which are being introduced to have internet connectivity in or near the crop field so that these sensors can effectively work in that environment and ease the farmers in many ways.
2. It is expected that with the integration of such advanced technology with agriculture the quality of the production gets improved regularly and the ways of earning more increases for the farmers.
3. It is seen that just manual intervention in agriculture can result in increase in wastage of the resources and the crop produced so to reduce the unnecessary factors these technologies might help and as a result of it automated options might increase more in the near future.
4. With the help of predictive learning farmers can become smart and start committing to the technologies as it will provide regular information and solutions to the problems which can revolutionaries the whole idea of agriculture.
5. More opportunities of better and improved quality crops might arise which can convince the farmers about the effective approach of technology and the cost-efficient way is better than wastage of energy, time, and money unnecessarily.

Agriculture being the oldest yet the most important sector of industry and human profession is now being digitized with an aim of improving not only the agricultural machinery with massive use of technology such as AI, IOT, machine learning, etc. Moreover, advancement in technology has also contributed in boosting the country's economy and generating huge profits as well. When it comes to working on the whole system, various applications and software are developed. Therefore, the efficiency of such software becomes a huge responsibility. Therefore, various quality and efficiency tests are performed which make the agriculture industry a bit complex.

The application of electronic devices in the field of agriculture is increasing day by day. With the massive utilization comes multiple tests and evaluations. The software that is mainly used is complex. Therefore, the scope of testing has expanded. In order to achieve sustainable agriculture and improve functionality and maintenance of the machinery, some tests need to be implemented. Also, there is a huge amount of work required for testing as well. With a wider range of performance of machinery comes more complex software. Therefore, in order to meet all the requirements, overcome the adversity of tests being performed and set the needed specifications for software testing, a hardware-in-the-loop simulator (HILS) is widely being used.

Hardware in the loop simulation testing is a technique of dissembling sensors and mechanical components in a way that all the I/O are tested at an early stage before actual integration of the whole system. It is used in automotive expansion. There is a broad range of conditions under which the agricultural software and machinery that needs to be tested such as the operational settings, safety, and security and the instances of testing is quite difficult to render in a real time scenario. Therefore, here the role of hardware-in-the-loop simulator comes into play. This simulator can be flexibly used in setting the required conditions of the software very safely.

Moreover, it helps in overcoming the difficulty in the execution of the testing means if the software bugs are once identified during the end season of harvesting, then the bugs will not be fixed until the harvest season gets over. Therefore, by using hardware-in-the-loop simulator, the software testing can be proceeded at any time. Also, the hardware-in-the-loop simulator is a tool which is broadly used in automotive expansion of testing procedures which makes it a way easier for the people to conduct various tests in less time. Applications of agricultural machinery where the use of HILS is applied:

1. Robot Tractor: Since the driverless tractor robots work using the satellite signals or base stations to know the exact location and its route on a computer tablet. In order to know how these tractors, work in case of any type of fault or failure, HILS is used to ensure the safe determination and handling of such difficulties.
2. Combine Harvester: These automated harvesters use sensors in order to detect the grain loss or wastage during sorting processes and therefore, automatically controls the amount of loss. Using HLS provides a sort of advantage in controlling the loss in non-harvesting seasons. Therefore, it becomes much easier to verify efficiency and quality of the operations being performed.

4 Testing

Testing basically refers to an interconnection between machine behaviour with defined standards under certain conditions. This is one of the major processes of any project development. It is mainly required in order to ensure the proper working of the system in accordance with some crucial local requirements. Testing in the field of agriculture is equally important as the usage, be it agricultural machinery testing, soil testing, engine, or pumps testing. These test results help not only farmers but customers as well in order to ensure the machine performance.

Moreover, it provides the facility of selection of the best suitable machinery in accordance with certain conditions resulting in high yield and popularization. It also encourages the quality as well as functionality stability. Manufacturers are also somehow benefited as proportionate data for similar machines is provided to them which in turn improve their product design. The farm machinery testing requires the determination of the:

1. Working features of machinery/equipment.
2. Strength requirement.
3. Endurance.
4. Wear and tear testing of tools.
5. Pressure on machine due to loading.

Therefore, proper strategizing and heedful execution are utmost important. The tests should be designed in such a feasible way so that the equipment ensures durability during the tests in labs and fields as well. There are mainly two tests required for evaluating the performance of agricultural machinery.

1. Commercial: Commercial tests are for demonstrating the characteristics of equipment that are commercial production ready.
2. Confidential: Confidential tests are basically for providing confidential details of the performance to commercial production that may be required by the applicant.

4.1 Software Testing in Agriculture

There are different types of testing a software that may be applicable while performing software testing as discussed below:

1. *Functional Testing:*
   There are various agricultural applications and software that connect a lot of users in bulk from farmers to buyers and from manufacturers to vendors. It allows the organizations/companies to check the applications (apps) over internet, mobile, and computer platforms. Functional testing not only improves the user experience but also removes the errors or bugs and mends them accordingly and releases the output data effectively.

2. *Performance Testing:*

It is another testing that estimates the speed, readiness, stability, and responsiveness of a software under the workload. Performance testing helps in building an app which is free of errors or bugs. Moreover, it involves comparison of multiple systems/configurations in order to ensure improved performance and ensures the best suitable configuration for the application.

3. *User Experience Testing:*

This testing is also known as usability testing. It measures at what level the software is user friendly. A small set of target audiences use the application and identify the usability defects if any. It mainly stresses on ease of using the software. Usability testing ensures greater efficiency and target audience satisfaction and reliability.

4. *Compatibility Testing:*

Since the use of technology in the field of agriculture is increasing day by day; therefore, various agricultural software are now being made. In order to ensure the efficient working of the software on all the platforms, compatibility testing is performed. The applications are designed in such a way that it can be operated on different browsers, databases, OS, hardware, devices, and network. It is basically a non-functional testing which ensures the customer needs and demands are satisfied. The application needs to undergo the compatibility testing in order to remove all the possible failures that could occur due to different versions, internet speed, etc.

5. *Regression Testing:*

When it comes to the agricultural industry, regression testing plays a key role in verification of its working and performing various modifications and updates in the form of bug fixes, functionality improvement needed if any without changing the functionality of the software.

5 Test Standards in Agriculture

Several different organizations and foundations have been established by different countries that aim at testing various farm machinery that keeps a check on all the sampling procedures being executed. These organizations firstly generate unique standard test codes for different types of machines which acts as a ground for testing.

Examples of such standards are as follows:

1. Bureau of Indian Standards: Indian government has established the Bureau of Indian Standards that will provide the test code for equipment such as tractors, tillers, crop production machinery, etc.
2. Nebraska Testing Centre in USA.
3. British Standards Institutions London, in UK.
4. Organization for the Economic Co-operation & Development (O.E.C.D.)
5. Paris and International Organization for Standardization (ISO).

6 The International Organization for Standardization (ISO)

ISO stands for "The International Organization for Standardization". ISO is a non-government and independent organization. It is an organization established at Geneva, Switzerland in 1947. This organization is structured in order to facilitate the commuting of goods and services internationally. It also builds cooperation in technical, technological, scientific, and economic fields as well. It also provides a common platform for developing and testing tools with a global network across the world having members from different countries [27–31].

In agriculture, the productive utilization of land is the most important factor for sustainable development. Therefore, test standards play a key role in providing the necessary information and tools for use of machinery and farming methods. ISO for agriculture encompasses each and every part of farming from irrigation to farm management and ensuring the proper supply of quality products in the food chain, thereby contributing to a better future. The following are the domains that ISO standards have for agriculture:

1. Tractors and other equipment for agriculture.
2. Fertilizers and soil conditioners.
3. Animal feeding equipment.
4. Defensive clothing.
5. Agricultural electronics.
6. Food products and safety.

6.1 Case Study

As we have already discussed about AIIoT and test standards of agriculture we know about the basic steps which be seen as a field of cooperation between AI and IoT. In this section we will be discussing the various real-life examples which are being used in the world today and their ideology behind the mechanism which they are using to provide sustainable agriculture and good quality farming according to the agricultural test standards [32–36].

1. *DTAC telecommunication and NECTEC, Thailand:*
   The telecommunication firm named DTAC which is fully owned by the Telenor group began a precision agriculture solution. They joined hands with the National Electronics and Computer Technology Centre of Thailand and the Department of Agricultural Extension to initiate a smart farming solution. Within this solution, the brand aims to provide the farmers with the required help for the sake of good farming by using the technology.
   They initiated a pilot plan in which the IoT technology works to observe, interpret, and predict the circumstances concerning the products of the

agricultural firm so that the test standards can be retained and the appropriate product is produced. This innovative solution helps the farmers by providing the path to a more specific farming method that can improve crop yields, inspect the status of stocks, and decrease production expenses. DTAC takes care of the wireless connectivity of the internet along with cloud computing, NECTEC works to develop analysis and sensory systems to inspect for signs of farming criteria.

2. *Vodafone, New Zealand:*

In New Zealand, the Vodafone company tries to enhance production and decrease expenses by precision farming tools to help the farmers so that they can produce products under the guidance of the international test standards. This solution enables farmers to predict the amount of fertilizer the soil needs to grow the crops. GPS agriculture machine is placed in the carrier machine dispersing the fertilizer and information is transmitted using the network of Vodafone company to provide a secure service for Precision Farming. The data received is then loaded on a digital map to know the amount of fertilizer which is required in the field and at which area.

With detailed information about the land and the fertilizer used, farmers can directly find unspecified wastage and can diminish the wastage of resources. Vodafone's agriculture system solely relies on GPS machinery. The operator provides the unique SIMs which are needed for the smooth flow of data.

3. *Libelium, Colombia:*

Red Technopreneur Colombia has stationed wireless sensors networks with the help of Libelium technology to observe the crop yield in the area of Lembo which is located in the Santa Rosa de Cabal. The plantain crops have been observed by various sensors with the help of Waspmote Plug & Sense. The sensors enable farmers to keep a watch over the following factors which include soil moisture, temperature, humidity, soil temperature, fruit diameter, and solar radiation. Monitoring of the above-mentioned parameters is the ultimate aim of the company so that the product is fit to use. This initiative also helps the farmers to produce the product by following the international test standards of agriculture. This also helps the farmers to predict and understand the harvesting projection, water usage, diseases, fertilizers consumption, and to understand which nutrients and minerals are required by the soil for a better yield.

7 Conclusion

Agriculture self-regulation and test standards of agriculture are the subjects which require people's attention as they are related to human sustenance. The earth's population is escalating at a very fast pace and the result of such expansion in population means a requirement for more food. Now the food that is produced needs to be fit for consumption so the process of production of crops needs to follow some rules and regulations to attain the certification to prove the authenticity of the product. Conventional techniques practised by producers are not enough to attend the growing market so the need to switch from the conventional method to the modern methods is required. The implementation of technologies such as Artificial Intelligence and Internet of Things has given a new shape to the area of agriculture. These technologies can not only produce a good yield but also can make sure the implementation of the rules and regulations. These test standards of agriculture can ensure that the crop is produced in a safe environment. This all is only possible if the combination of AI and IoT is trained in such a way that they can monitor the test standards of agriculture in the fields. Thus, AI and IoT can be seen as a platform to provide humanity with the right harvest and AIIoT serves a major role in test standards of agriculture.

References

1. Padikkapparambil, J., Ncube, C., Singh, K. K., & Singh, A. (2020). Internet of Things technologies for elderly health-care applications. In *Emergence of pharmaceutical industry growth with industrial IoT approach* (pp. 217–243).
2. Singh, M., Sachan, S., Singh, A., & Singh, K. K. (2020). Internet of Things in pharma industry: possibilities and challenges. In *Emergence of pharmaceutical industry growth with industrial IoT approach* (pp. 195–216).
3. Atzori, L., Lera, A., & Morabito, G. (2010). The internet of things: A survey. *Computer Networks, 54*, 2787–2805.
4. Koreshoff, T., Robertson, T., & Leong, T. (2013). Internet of Things: A Review of Literature and Products. In: *OzCHI'13 Proceedings of the 25th Australian Computer-Human Interaction Conference: Augmentation, Application, Innovation, Collaboration* (pp. 335–344). (ACM, Adelaide, Australia).
5. Li, S., Xu, L., & Zhao, S. (2015). The internet of things: A survey. *Information Systems Frontiers., 17*, 243–259.
6. Whitmore, A., Agarwal, A., & Xu, L. (2015). The internet of things - a survey of topics and trends. *Information Systems Frontiers., 17*(2), 261–274.
7. Sarma, S., Brock, D. L., & Ashton, K. (2000). *White paper: The networked physical world: Proposals for engineering the next generation of computing.* Commerce & Automatic-Identification.
8. Ibarra-Esquer, J., González-Navarro, F., Flores-Rios, B., Burtseva, L., & Astorga-Vargas, M. (2017). Tracking the evolution of the internet of things concept across different application domains. *Sensors, 17*(6), 1379. 1–24.
9. Nayyar, A., Puri, V., & Le, D. N. (2017). Internet of nano things (IoNT): Next evolutionary step in nanotechnology. *Nanoscience and Nanotechnology, 7*(1), 4–8.
10. Solanki, A., & Nayyar, A. (2019). Green internet of things (G-IoT): ICT technologies, principles, applications, projects, and challenges. In *Handbook of research on big data and the IoT* (pp. 379–405). IGI Global.
11. Batth, R. S., Nayyar, A., & Nagpal, A. (2018). Internet of robotic things: driving intelligent robotics of future-concept, architecture, applications and technologies. In *4th International Conference on Computing Sciences (ICCS)* (pp. 151–160).
12. Comfort, L. K. (2019). *The dynamics of risk: Changing technologies, complex systems, and collective action in seismic policy.* Princeton University Press.
13. Creedy, G. D. (2011). Quantitative risk assessment: How realistic are those frequency assumptions? *Journal of Loss Prevention in the Process Industries, 24*, 203–207.

14. De Marchi, B., & Ravetz, J. R. (1999). Risk management and governance: A post-normal science approach. *Futures, 31*, 743–757.
15. Durga Rao, K., Gopika, V., Sanyasi Rao, V. V. S., Kushwaha, H. S., Verma, A. K., & Srividya, A. (2009). Dynamic fault tree analysis using Monte Carlo simulation in probabilistic safety assessment. *Reliability Engineering and System Safety, 94*, 872–883.
16. Goodfellow, I. J., Bengio, Y., & Courville, A. (2006). *Deep learning*. The MIT Press.
17. Navulur, S., Sastry, A.S.C.S., Giri, M. N., & Prasad. (2017). Agricultural management through wireless sensors and Internet of Things. *International Journal of Electrical and Computer Engineering (IJECE), 7*, 3492–3499.
18. Sisinni, E., Saifullah, A., Han, S., Jennehag, U., & Gidlund, M. (2018). Industrial Internet ofThings: Challenges, opportunities, and directions. *IEEE Transactions on Industrial Informatics, 14*, 4724–4734.
19. Aggarwal, P. K., Jain, P., Mehta, R., Garg, R., Makar, K., & Choudhary, P. (2021). *Machine learning, data mining, big data analytics for 5G enabled IoT, Blockchain for 5G enabled IoT* (pp. 351–375). Springer.
20. Minerva, R., Biru, A., & Rotondi, D. (2015). Towards a definition of the Internet of Things (IoT). IEEE Internet of Things. 1–86.
21. Ashton, K. (2009). That 'internet of things' thing. *RFID Journal., 22*, 97–114.
22. Zhang, L., Dabipi, I. K., & Brown, W. L. 2018. Internet of Things applications for agriculture. In Q. Hassan (Ed.). *Internet of things A to Z: Technologies and applications*.
23. Lin, J., Yu, W., Zhang, N., Yang, X., Zhang, H., & Zhao, W. (2017). A survey on internet of things: Architecture, enabling technologies, security and privacy, and applications. *IEEE Internet of Things Journal, 4*, 1125–1142.
24. Zhang, X., Zhang, J., Li, L., Zhang, Y., & Yang, G. (2017). Monitoring Citrus soil moisture and nutrients using an IOT based system. *Sensors, 17*, 447.
25. Hicham, K., Ana, A., Otman, A., & Francisco, F. (2017). Characterization of near-GroundRadio Propagation Channel for wireless sensor network with application in smart agriculture. *In Proceedings of the 4th International Electronic Conference on Sensors and Application, SolelyOnline*.
26. Aggarwal, P.K., Sharma, S., Riya, Jain, P., & Anupam. (2021). Gaps identification for user experience for model driven engineering. In *Proceedings of the International Conference on Cloud Computing, Data Science & Engineering- Confluence*.
27. Aggarwal, P. K., Grover, P. S., & Ahuja, L. (2018). Exploring quality aspects of smart mobile phones applications. *Journal of Advanced Research in Dynamical and Control Systems (JARDCS), 11*, 292–297.
28. Aggarwal, P. K., Grover, P. S., & Ahuja, L. (2019). Assessing quality of Mobile applications based on a hybrid MCDM approach. *International Journal of Open Source Software and Processes (IJOSSP), 10*, 51–65.
29. Jain, P., Sharma, A., & Ahuja, L. (2018). The impact of agile software development process on the quality of software product. In *Proceedings of the International Conference on Reliability, Infocom Technologies and Optimization (Trends and Future Directions) (ICRITO)* (pp. 812–815).
30. Jain, P., & Sharma, S. (2019). Prioritizing factors used in designing of test cases: An ISM-MICMAC based analysis. In *Proceedings of International Conference on Issues and Challenges in Intelligent Computing Techniques* (ICICT).
31. Jain, P., Sharma, A., & Aggarwal, P. K. (2020). Key attributes for a quality mobile application. In *Proceedings of the International Conference on Cloud Computing, Data Science & Engineering Confluence* (pp. 50–54).
32. Elijah, O., Rahman, T. A., Orikumhi, I., Leow, C. Y., & Hindia, M. N. (2018). An overview of Internet of Things (IOT) and data analytics in agriculture: Benefits and challenges. *IEEE Internet of Things Journal, 5*, 3758–3773.
33. Jain, P., Singhal, A., Chawla, D., & Shrivastava, V. (2020). *Image recognition and segregation using image processing techniques. TEST Engineering and Management*.

34. Khanna, A., & Kaur, S. (2019). Evolution of Internet of Things (IOT) and its significant impact in the field of Precision agriculture. *Computers and Electronics in Agriculture, 157*, 218–231.
35. Jain, P., Aggarwal, P. K., Chaudhary, P., Makar, K., Mehta, J., & Garg, R. (2021). Convergence of IoT and CPS in robotics. In *Emergence of cyber physical systems and iot in smart automation and robotics* (pp. 15–30).
36. Thea, K., Martin, C., Jeffrey, M., Gerhard, E., Dimitrios, Z., Edward, M., & Jeremy, P. (2017). Food safety for food security: Relationship between global megatrends and developments in food safety. *Trends in Food Science & Technology, 68*, 160–175.
37. Jain, P., Anupam, Aggarwal, P. K., Makar, K., Shrivastava, V., & Maitrey, S. (2020). Machine learning for web development: A fusion. In *Proceedings of 2nd International Conference on AI and Speech Technology.*

Study and Analysis of 5G Enabling Technologies, Their Feasibility and the Development of the Internet of Things

Rubaid Ashfaq

1 Introduction

From the beginning, people have communicated, share and exchange ideas. Throughout history, these forms of communication have evolved according to different needs, which have generated the improvement in information technologies and communication (ICTs). At present, distance communication constitutes one of the most significant advances. Telecommunications provide facilities for exchange and access to information, from one point to another. Thanks to this, remote communications have constantly been advancing. In 1876, Alexander Graham Bell patented the telephone, thereby which the human voice could be transmitted and received for the first time. From this fact, the telephone was evolving in parallel with technology. It is here where the ability to transmit data at high frequencies began to develop [1].

Over time, standards and technologies have been adapted, established by international regulatory bodies and entities telecommunications to access services and interconnect facilities provided by different manufacturers. Currently, mobile telephony is a means of personal communication, which allows access to various cloud services, such as online stores, web portals, digital magazines, blogs, entertainment pages, employment platforms, social networks, among others, at relatively high transmission rates.

However, the demand for communication networks is higher and higher due to the constant proliferation of new terminal equipment and users. The situation leads to an overload of traffic and a possible collapse in the network. That is why the telecommunications industry is obliged to develop new forms and methods to expand and revolutionize communication networks.

R. Ashfaq (✉)
Amity School of Communication, Amity University, Noida, India

© The Author(s), under exclusive license to Springer Nature Switzerland AG 2021
F. Al-Turjman et al. (eds.), *Intelligence of Things: AI-IoT Based Critical-Applications and Innovations*,
https://doi.org/10.1007/978-3-030-82800-4_5

With the evolution of mobile networks, utilities, applications, and devices are more and more abundant. The high transmission capacity data allows you to navigate without problems in the web of networks, internet, which allows there to be fluid data traffic that allows the management of a lot of information. Today, human beings have taken a great step in technology. Currently, devices such as televisions, telephones, and tablets can access the internet. However, these objects still need the manipulation of the human being to make decisions. From this point, the Internet of Things was born (Internet of Things—IoT). The connectivity between things, that is, between devices with the internet, facilitates the automatic exchange of information with other devices or control centers without human intervention. This technology can collect large amounts of data, benefiting in control, monitoring, automation, and operation, creating development and evolution opportunities for users, companies, and cities. But for this to be implemented, need networks with improved spectral and energy efficiency. It is also important that the delay between the transmitting and receiving equipment is imperceptible, which is why the fifth-generation 5G networks play a great role in developing this new technology.

2 Background

On January 30, 1962, Nikolas Tesla said in an interview for Collier's magazine: "When wireless connection is perfectly applied to the whole Earth, this will become one big brain, which it is, all things are particles of a real and rhythmic whole. We can communicate between us instantly, regardless of distance. Not only this, but through television and telephony, we will see and listen to each other as perfectly as if we were face to face. Despite the distances intermediate thousands of thousands, and the instruments through which we will be able to do this, they will be incredibly simple compared to our current phone. A man can carry one in his pocket." [2].

These were the words that marked the foundation of the world internet. In 1969, the ARPA (Advanced Research Projects Agency) first designed and launched the ARPANET, the predecessor network of what is currently known as the internet. This network would be the beginning of the development of global interconnection, which would form the entire planet into a brain, as stated by Tesla in the interview. The whole world began to evolve in technology constantly, the protocols NCP, TCP/IP, UDP, among others, were created. Kevin Ashton gave know for the first time the term "Internet of things" at a convention in Procter & Gamble, where he presented the IoT concept.

The internet of things has evolved little by little through the years. However, it is not yet 100% adapted. In 1990 John Romkey created the "Internet Toaster," which was, in other words, the first "thing" connected to the internet. The toaster was connected via a network TCP/IP and controlled by SNMP (Simple Network Management Protocol), which contained turning on and off the device. Wireless networks, whether WIFI or a cellular network, were undoubtedly a very important revolution for the internet world, one of the most relevant drivers for the interconnection of objects. Thanks to advances in wireless networks, the IoT has evolved little by little in recent years. Above all, the emergence of technologies such as WSN (Wireless Sensor Network) and M2M (Machine to Machine), the IoT has also been working in various sectors such as the industries and factories with sensor networks, which allow monitoring, control, and management of product development through M2M communication.

The IoT is not just a network that interconnects devices with each other. It is more that this is a new reality in which quantities are obtained exponentially high information in real-time, which facilitate the functionalities and applications that the human being executes daily, say the IoT is an intelligent and autonomous network, which is capable of managing the information without human intervention. Thanks to this, you can envision a great development in the economy since it generates opportunities and benefits for society.

3 Definition of the Problem

The transmission rates and high latency of today's mobile networks are not effective enough to fully adopt the internet of things. This is when the 5G geared plays a predominant role because this is the ideal platform for these millions of devices to connect to the internet at an optimal capacity where all these terminals can manage the information since they offer thousands of times more traffic than current networks and will be ten times faster than networks 4G and 4.5G, and an infinitesimally short response time.

4 Objectives of the Research Problem

4.1 General Objective

Analyze the technologies of a 5G mobile broadband network and its feasibilities to promote the Internet of Things (IoT).

4.2 Specific Objectives

- Understand essential technologies of a 5G network.
- Analyze and understand the operation of the IoT.
- Relate advantages, feasibilities, and compatibility between 5G and IoT.
- Publicize the security and protection of these networks.
- Present the possible solutions for the current IoT problem.

5 Hypothesis

A study of the technologies that will be used for future networks is carried out generation, which can provide solutions to the problem of connectivity and response of the Internet of Things (IoT) because this mobile network broadband has high transmission rates at low latency, which favors the IoT not only to facilitate functions and utilities to beings humans, is also intended to revolutionize the world, it says a virtual reality where objects connected to the internet will take decisions, which will meet the needs of human beings, promoting a more comfortable and timely environment.

6 Research Methodology

In this research work, the descriptive methodology is used and theoretical since a study of the mobile broadband network of fifth generation to improve and enhance things. By analyzing various tasks and tests carried out by scientific entities and recognized universities, the characteristics and basic operations of a 5G network related to the internet highlight its advantages and improvements for this new technology.

7 Evolution of Mobile Networks from 0G to 5G: Basic Mobile Broadband Concept

Mobile telephony was born from the concept of radio access in which each user is given a radio wave frequency, in which he can operate. The mobile network is based on the constant location of users, through a network infrastructure composed of transmitters and receivers better known as Base Station (BS), offering coverage to a specific radio, the coverage where the service is provided is known as a cell, from which it takes the name "cellular network" since it is composed of a variety of cells, being transparent to the user in which cell is found [3].

The proper functioning of these networks depends on two systems:

- Paging: User's specific location system.
- Handover: The process in which the user moves from one cell to another without losing the connection, that is, the handover from one node to another. These mobile networks have a basic network infrastructure, where their design is shown below in Fig. 1.

The mobile broadband or mobile phone network works similar to fixed telephony, interconnecting two points through the network modules of a service provider or operator, which is committed to the management and delivery of the service.

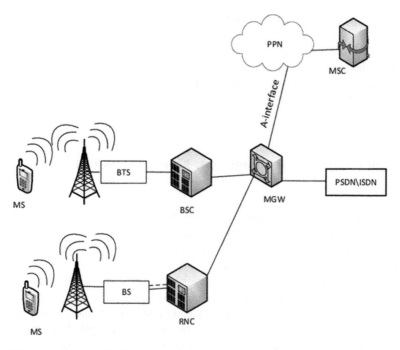

Fig. 1 Base Architecture of a Cellular Network

Table 1 Long Term Evolution Category. Source: Prepared by the author

Long Term Evolution Category	Velocity	Carrier option added
Category No 12	600 Mbps download	3 × 20 MHz downstream
	100 Mbps upload	2 × 20 MHz upload
Category No 10	450 Mbps download	2 × 20 MHz downstream
	100 Mbps upload	2 × 20 MHz upload
Category No 9	450 Mbps download	3 × 20 MHz downstream
	50 Mbps upload	
Category No 7	300 Mbps download	2 × 20 MHz downstream
	100 Mbps upload	2 × 20 MHz upload
Category No 4	150 Mbps download	2 × 10 MHz downstream
	50 Mbps upload	

However, in contrast with landline telephony, electrical radio wave transmissions are in charge to link these points. Terminal equipment or UE user equipment, through the air interface, establishes a communication with the nearest antenna or base station (BTS). This communicates with the base station controller (BSC); usually, a BSC has dozens or even hundreds of BTS under its control. The BSC handles radio channel assignment, receives measurements from phone mobiles, and controls the handover from BTS to BTS. The Mobile Switching Center (MSC) is responsible for routing the various communications, either to a fixed or mobile network. To provide good QoS (quality of service) in touch, the UE must be within the BTS area can offer coverage. The range of this coverage is limited. This area is known as a cell.

To offer the broadest possible coverage, the service providers install hundreds and even thousands of cells to cover in its entirety a determined area of land. That is, there are no spaces between cells so that the user can have communication uninterrupted.

8 Mobile Broadband Evolution: First Generations

In the late 1940s, the Motorola company developed the zero generation (0 g), which originated during WWII. This technology allowed communication over long distances between war troops. These telephones were known as radiotelephones, AND they worked in the bands of HF AND VHF frequency.

The 0G generation used radio systems, initially implemented amplitude modulation. However, because these systems are prone to noise, frequency modulation was implemented, and therefore better audio quality could be obtained. The American company Bell was the pioneer of these systems, which were highly large and heavy. These were installed inside the vehicles. Implementing this generation's requirements was considerably expensive; however, it was operating with specific updates that emerged from 1946 to the beginning of the 80s.

This generation used ARP (Autoradiopuhelin) technology. Finnish commercial telephony launched in 1971 had coverage of 100% in Finland; however, as time went by, to become congested, for which it was replaced by modern technology, began being half-duplex and later full-duplex, that is, it allowed a two-way voice transmission. This generation zero marked by an analogue signal did not encrypt the calls, so it left a lot to be desired in terms of security, and also did not support the fundamental characteristic of cellular networks current that is the handover, for which the calls in this generation are they cut automatically. In the late 1970s and early 1980s, several cellular communication systems were characterized by transmitting voice analogue by frequency modulation.

These first systems or also known as standards were the following:

- AMPS: Introduced in the late 1970s in the USA. (Advance Mobile Phone System).
- NMT: Considered the first 1G standard. (Nordic Mobile Telephones System).
- TACS: (Total Access Communication System).

These first-generation networks offered good voice quality. However, the spectral efficiency was minimal, which led to the congestion; another significant limitation that this network had was managing network control messages such as handover. These were carried over the voice causing interference and noisy sounds. The low capacity of transfers between cells was due to FDMA (Multiple Access by Frequency Division.) Because its name divides or splits the bandwidth into several segments, which reduce the transfer capacity and even more if the traffic is very high [4].

9 Digital Generation

The second-generation or better known as the 2G network was characterized by integrating digital systems, where not only voice was transmitted, data could also be sent at higher speeds. This technology emerged in 1982 as GSM (Global System for Mobile Communications), which had a technological success because this generation manages to separate the transport layer and the control layer.

This new system's increase in traffic generated several vital points to improve, such as spectral efficiency, data, coverage, and capacity. The process that must be carried out to improve traffic is implementing base stations' additional lower powers, thus allowing reuse of frequencies at shorter distances and thus achieving progress in the spectral efficiency since it will enable access to more users Mhz.

GSM brought one of the essential services was the Short Messages (SMS), which allowed sending of 160 characters. These services provided by GSM were undoubtedly a very noticeable advance in the communications. 2 G data rates were 14 kb/s up to 22 kb/s were relatively acceptable speeds. However, the constant increase in user demand made GSM will evolve, improving its services. GPRS (General Packet Radio Service) is a data service that reuses the GSM network infrastructure, with some changes, such as adding a new layer of control of access to the medium MAC and one of control of RLC radio link. With this new service, it is possible to reach speeds of up to 171 kb/s. These speeds can be adapted and optimized depending on canal conditions. This GSM extension had much success among the operators to adjust their networks to this new technology [5].

Then GSM evolved under the name EDGE (Enhanced Data for GSM Evolution). This improvement requires a new radio interface. This evolution improves the capacity and coverage of GSM in places where traffic is cumbersome. However, it was not very successful because its implementation required a restructuring of the network infrastructure, completely modifying the radio interface. Therefore, its installation had a much higher cost than GPRS.

10 Multiple Accesses by Code Division

CDMA, a technology was already present in 2G networks; however, this was the base technology on which the future generation 3G would work and later evolutionary to W-CDMA. This technology was designed by the Qualcomm company and was proposed in 1993, and later it was standardized as IS-95 for second-generation networks.

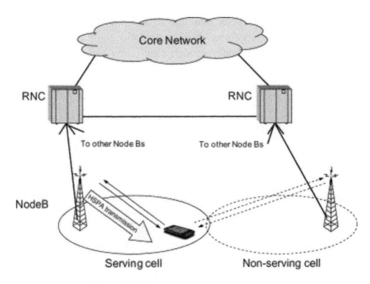

Fig. 2 Multiple Access by Code Division

This technology provides better quality, capacity, and coverage, which facilitates the simultaneous communication of multiple users in the same medium, thus improving spectral efficiency. CDMA is a technique in which data orbits are modulated by a high-frequency orthogonal sequence known as random pseudocode, as shown in Fig. 2. This signal or code sticks to the bits that you want to send through an XOR logic gate; thus, the resulting sequence is the sequence to be transmitted. Multiple signals from different users are transmitted on the same frequency band; therefore, it must have the same code for the receiver to recover the signal, which will be multiplied by the received signal. This makes CDMA very safe and robust as it does not need guard bits [6].

11 Third Generation

The high demand generated by users in terms of more incredible speed, better capacity and coverage, and the integration of new services forced the mobile phone industry to evolve in its systems. UMTS (Universal Mobile Telecommunication Systems) is the first third generation used W-CDMA technology and was standardized by the 3GPP group (Third Generation Partnership Project). Multimedia Services as video conferences is one of the advantages achieved with third generation, thanks to its increase in transmission rates with 384 kb/s.

UMTS was characterized by using CDMA; however, the applications and services of a different nature can be achieved with UMTS, made technologies such as W-CDMA (Wideband Code Division Multiple Access) and TD-CDMA (combination of TDMA and CDMA). How I know, see in Fig. 3 the UMTS access network

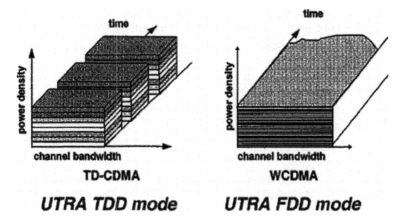

Fig. 3 Services and applications of W-CDMA and TD-CDMA technologies. Source: Adell Hernani and Telefónica [3]

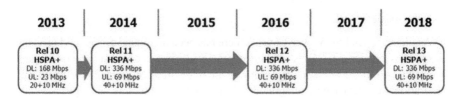

Fig. 4 HSPA + Evolution

(UTRA) uses duplexing by Frequency Division (FDD) over WCDMA for services and situations where traffic is symmetrical.

TD-CDMA uses duplexing by time division for public cells and wireless services where the traffic is denser [3]. 3G is essentially a continuation of 2G. It provided more coverage and capacity; however, one of the characteristics that revolutionized this generation is the integration of IP services. Initially, the 3G network was ATM, later with the integration of the internet to the mobile phone,

New applications emerged, such as:

- Mobile banking.
- E-commerce.
- Mobile operating systems.
- Dedicated software and applications.
- Location.

This technology allows access to services and applications to even higher speeds, soon after it evolved again, to HSPA + As can be seen in Fig. 4, HSPA + has had an evolution of about.

Simultaneously, in 2013, a download speed of 168 Mbps was obtained and an upload speed of 23 Mbps. In 2014, the Release was standardized 11 a version even faster than the previous one providing speeds of 336 Mbps download and 69 Mbps upload. It should be noted that at present, LTE technology. However, the operators can access 2G, 3G technologies and their evolutions, depending on terminal equipment capacity.

12 Long Term Evolution (LTE)

. 4G emerged as an evolution of 3G, through the development of new services available for mobile devices. The three most important factors that made 4G evolve are as follows:

- Greater speed and capacity.
- New IP services, such as VoIP, IPTV, instant messaging on IP, among others here the network worked all-IP.
- Network optimization.

As illustrated in Fig. 5, over 13 years there has been an impact on the market as far as speed is concerned, 3G as explained previously has evolved in several stages, from 2003 until 2009 each stage with an increase in speeds. In 2010 the fourth generation (4G LTE) was integrated offering a maximum rate of approximately 150 Mbps, however, as in all previous generations, goes through different phases, which is why LTE evolved to LTE-Advanced (LTE-A) offering speeds of up to 1 Gbps theoretically.

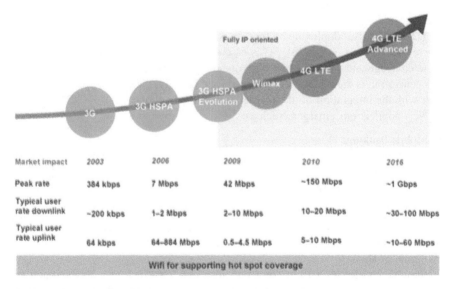

Fig. 5 Fourth-Generation Speeds

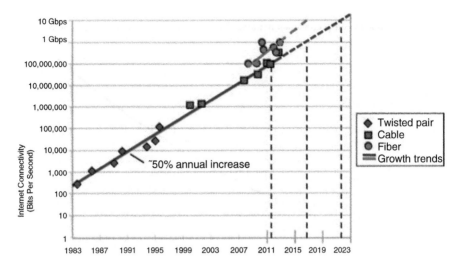

Fig. 6 Carrier aggregation

These speeds are theoretical, so in practice, they depend on many factors, such as congestion, conditions Climate, Network Policies, Licensed Spectrums, Backhaul, etc. However, these affectations may not be very noticeable since it is very unusual to find applications that need to operate at a speed of 1 Gbps.

When it comes to capacity, it is a very important factor for the development of any mobile technology. The increase in DATA that has occurred overtime has been very demanding, as can be seen in Fig. 6 from 1985 to 2025 the volume of DATA per month has increased to 1Gbs, an exponentially high amount. Most of the data growth is attributed to video-based services. Therefore, it is not only about increasing the speed, but also the ability to get more information and more users.

One of the key techniques for success in terms of speed and capacity of this fourth generation is carrier aggregation (Carrier Aggregation—CA). In Fig. 7 this process can be observed, which consists of the aggregation of multiple carriers of approximately 20 MHz or less. That is, it allows the operation of several fragments of the spectrum, towards the same user or terminal equipment UE. This technique allows the combination of the signals of the different carriers so that it is possible to achieve a bandwidth of up to 100 MHz.

As can be seen in Table 2.1 4G as well as its predecessors has had different phases, through which it can offer greater benefits to the user. In most countries, it is already standardized the fourth generation, and day by day, they are adhering to new devices and terminal equipment, which will undoubtedly make the future the home for fifth-generation networks and the IoT [7].

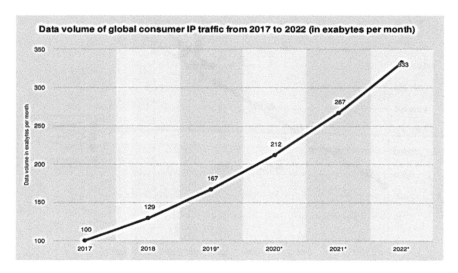

Fig. 7 Prolific volume increase diagram of DATA

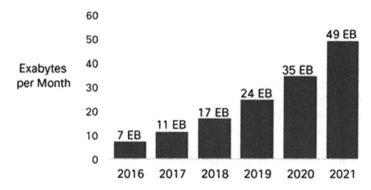

Fig. 8 Statistics of Increment of DATA

13 Essential Technologies for 5G Network Deployment

Fifth-generation networks (5G) are the set of new technologies, which when integrated offer new opportunities and applications in various sectors. This new generation plans in 2021, manage 1000× more capacity, that is, approximately 49 EB/month, according to statistics as shown in Fig. 8.

With the constant growth of users or devices that connect, to the internet, and at the same time process information from large quantities, several key points must be considered, Fig. 9 shows the requirements of the fifth-generation network in contrast to 4G, these key capabilities were established within the IMT-2020 standard, by the ITU (International Telecommunication Union). In 4G transmission rates reach 1 Gbps while the fifth generation proposes connectivity with a transmission rate of

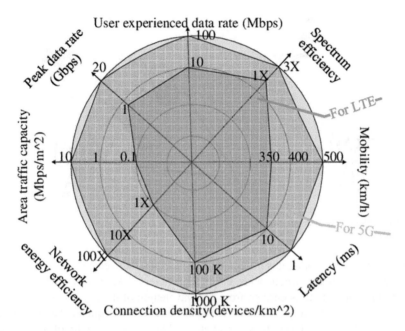

Fig. 9 Requirements for key capabilities. Source: Mumtaz et al. [8]

10 Gbps and 20 Gbps theoretically, for scenarios ideals within fifth-generation networks, the number of users and terminal equipment will have increased, which is why 5G promises provide transmission rates of 100 Mbps for dense areas and 1 Gbps for isolated areas.

Mobility is a key point in 5G, offering a very high QoS, in the case of obtaining fluid connectivity, for UE with mobility of 500 km/h while 4G supports 350 km/h. This generation provides a density of up to one million devices per km^2, at a latency of 1 ms, intending to create new communication systems such as the IoT.

5G is not just a simple bandwidth upgrade like its predecessors, it is a new form of connectivity, functioning as a network virtualized in the cloud, where its ubiquitous connectivity, both for people and things, develop new business methods and platforms, for a useful development of the human being. For the success of the 5G network and to justify the points specified above, there are new technologies, which are a key to the development and operation of the fifth-generation networks, which are explained below.

13.1 Millimeter Waves

Fourth-generation networks work in a frequency range from approximately 800–2600 MHz, as new terminals are integrated into the network, traffic becomes heavy causing congestion and collapse of the network, therefore, in the

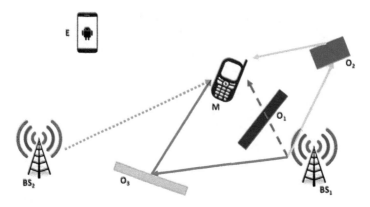

Fig. 10 Cellular networks current. Source: Researchgate

fifth-generation networks, it is necessary to expand the radio electric spectrum using a range of much higher frequencies, these frequencies are known as waves millimeter, the millimeter term is derived from the wavelength of the radio signals [8].

As can be seen in Fig. 10, cellular networks current, use frequency bands ranging from 300 MHz to 3 GHz, these frequencies have a wavelength of $\lambda = 1$ ma $\lambda = 100$ mm, on the other hand, it can be seen that the frequencies that go from 30 GHz to 300 GHz they have a wavelength $\lambda = 10$ mm and $\lambda = 1$ mm, respectively, hence its name "millimeter," these waves are known to operate in the EHF frequency band. These waves constitute a large part of the radio electric spectrum, currently, the wireless communication standard IEEE 802.11ad or better Known as WiGig, works in the 60 GHz frequency band.

Without, however, it is an unexplored space for mobile applications, therefore, it is very convenient to use this radioelectric space to satisfy the requirements of the fifth-generation mobile network, because, thanks to its nature allow very high transmission rates, a wide spectrum, and a large bandwidth which helps us avoid interference. Fig. 11 represents the frequency bands that have been assigned for fifth-generation mobile networks these are the bands.

From 26 to 28 GHz and 66 to 71 GHz, although studies are already being carried out within from the bands 41 to 43.5 GHz, to further expand the radio spectrum. One of the most prevalent advantages of millimeter waves for a 5G network is its wavelength is very small, therefore it suffers a high attenuation, this can be seen as a problem, which can be solved with the implementation of small cells; however, as there is a lot of space free, these high frequencies can be reused in shorter distances, thus improving the quality and coverage of the service [8].

However, not everything is advantages and benefits, the high frequencies that range between 57 and 64 GHz, 164 and 200, and 280 and 325 GHz, have characteristics not very favorable for the good performance of a 5G network since these frequencies correspond to the absorption of oxygen molecules and water vapor

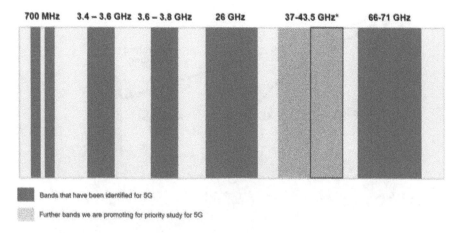

Fig. 11 Electromagnetic spectrum and its frequency bands

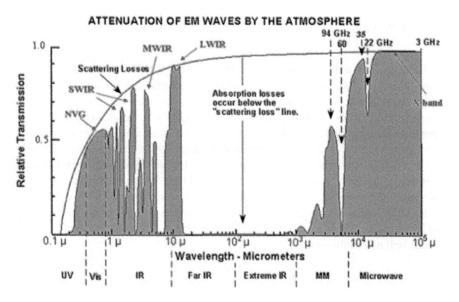

Fig. 12 Frequency assignment

absorption. I know you can see in Fig. 12. That in these instances there is a very high atmospheric attenuation.

The attenuation in the other available frequencies is very low so these frequencies are ideal for 5G, even so, there are factors that in general, can cause attenuation as is, the space that exists between the transmitter and receiver, rain, fog, dust, external interference, trees, buildings, among others. It is very important to consider the obstruction of obstacles, since to this problem there are some solutions, one of them is the configuration of the link.

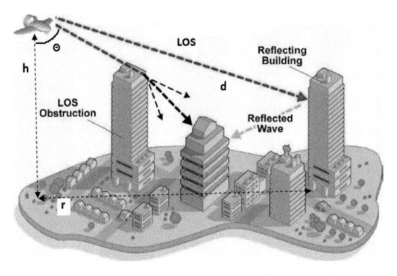

Fig. 13 LOS lock

There are two configurations, direct line of sight (LOS) and non-line of sight Direct vision,(NLOS). LOS is the ideal setting for communication links, as its name says it has a line of sight free of obstacles, that is to say, that the link will not have any loss. NLOS is an alternative when there is a LOS blockage, that is, in the line of sight between the transmitter and the receiver there is an obstacle obstructing the signal path, this is not the best solution for millimeter waves, since even though there will be a link between the sender and the receiver, this will have losses in its potency.

In Fig. 13 it can be seen that the UE has a LOS lock, which establishes an NLOS link, where the signal emitted by the antenna transmitter, bounces off the concrete of the building producing a reflection of the signal towards the UE, having access to the communication, however, this signal will bring power losses, so the quality of the service will not be the best.

14 Heterogeneous Networks

In today's wired networks, one can easily achieve a speed of 100 Gbps, but as you know you cannot perform wiring to each terminal equipment. Therefore, wireless networks are the best option to provide connectivity, however, just because it is wireless already has a higher margin of error due to several factors such as external interference, environmental conditions, power of the transmitter, etc. Therefore, high demand for transmission rates is generated. Multi-Gigabit is a wireless heterogeneous network system composed of transmitting stations that vary according to their magnitude, whether size, power, cost, etc.

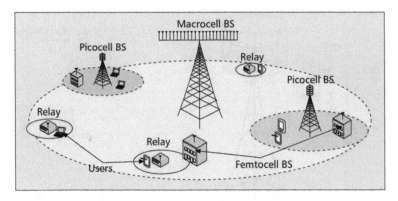

Fig. 14 HetNet with macrocells and small cells

Fig. 14 illustrates a basic scheme where in addition to the cells small relay nodes can also be implemented, which help to improve coverage in places where it is difficult to establish a cell small since these unlike the small cells operate without a backhaul cable connection.

15 Massive MIMO

So that wireless communications can have a link of point to point, it is necessary that at each end there is an antenna, be it transmitter or receiver, these antennas are responsible for transforming the electricity in an electromagnetic field (RF waves). Antennas or better known in the telecommunications world as AAU (Active Antenna Unit) are the main equipment of any wireless communication. To what throughout the evolution of cellular networks has been using different types of antennas with different powers, LTE has already implemented the technology MIMO (Multiple-input Multiple-output). Currently, 4.5G networks use multi-user MIMO technology. However, this technology is not scalable as it uses the same number of antennas and terminals.

Massive MIMO is a large-scale antenna system, this system helps to improve, spectral efficiency, in such a way as to improve the grid performance, by concentrating energy in spaces smaller, this, in turn, provides energy efficiency these systems require low energy consumption. Massive MIMO takes MU-MIMO to another level, the base stations of antennas with Massive MIMO technology are made up of an array of 100 to 200 antennas, to serve approximately 10–40 users within an instance of time as can be seen in Fig. 15. In other words, the number of antennas will be greater than the number of terminals.

When a terminal wishes to download information, or in the upload link; all terminals occupy frequency resources full time simultaneously. When establishing a link upstream the base station must retrieve the information of all terminals, while,

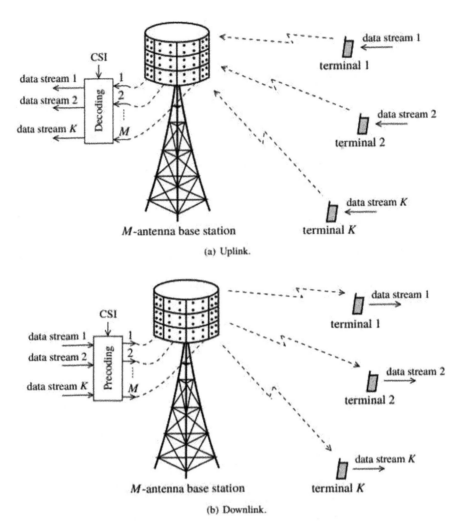

Fig. 15 Massive MIMO. Source: Marzetta et al. [9]

on the downlink, it has to ensure that the information reaches the desired terminal, through multiplexing processes and demultiplexed. Thanks to the huge number of antennas that massive MIMO, and the CSI (Canal Status Information), it is possible to get secure communication [9].

Another advantage of this technology is that it forms a beam with a small angular window, directly towards each terminal, therefore if you have more antennas, the beams will be smaller, so that will have more coverage for short ranges of distance, which needs less power consumption, which in turn provides uniformly a good simultaneous service to each terminal.

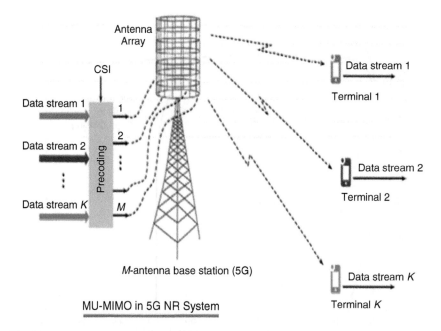

Fig. 16 Massive MIMO antenna array

Massive MIMO is viewed for the IoT (Internet of things), so which, it is planned to deploy this technology in different scenarios, to fit in different network infrastructures. As can be seen in Fig. 16, there are several forms of arrangement of antennas, to provide service in the various scenarios, such as arrays, of antennas, distributed in different structures, linear arrangement, cylindrical or flat. It should be emphasized that these arrangements will be fed through digital media whether optical or electrical. The potential that can be drawn from Massive MIMO, for implementation of fifth-generation networks, is very vast there are several points to consider that bring many benefits for the integration of new technologies:

- Increase in capacity.
- Cost reduction thanks to low consumption equipment.
- Robust system.
- Latency reduction.

With MIMO technology, it is possible to increase 10 times more, and concurrently, 100 times more energy efficiency is improved radiated. It is already known that massive MIMO consists of the operation of many antennas, which concentrate signals from transmission and reception in a small area, consequently, requires the use of a large number of radio frequencies, generally the same number of transmitting antennas, which becomes an unfavorable drawback for good performance.

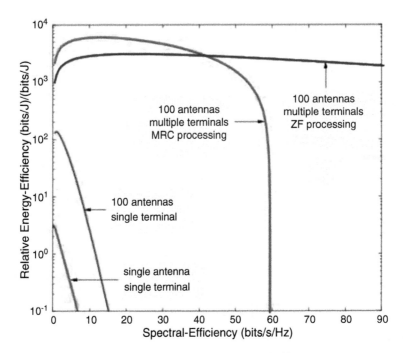

Fig. 17 Spectral efficiency gain diagram. Source: Marzetta et al. [9]

A solution to this problem is spatial multiplexing, where multiple antennas are used to transmit and receive and each channel carries independent information, increasing transmission rates.

This method can improve the efficiency of the link. As can be seen in Fig. 17 with the implementation of this technology, spectral efficiency can be improved, as can be seen in the diagram, if 100 antennas are used for a single terminal, you will not have the best spectral efficiency, because you are underusing the capacity that can be obtained 100 antennas for a single terminal, instead, it is implemented, 100 antennas for different users with spatial multiplexing, it is achieved get better spectral efficiency.

16 Beam Forming

Beam forming is an essential technique of transmission of data required for bulk MIMO to work as expected, this technology reduces the loss of signal propagation due to higher frequencies of millimeter waves.

In massive MIMO base stations, the algorithms of signal processing trace the best transmission path through the air for each user. They can then send individual data

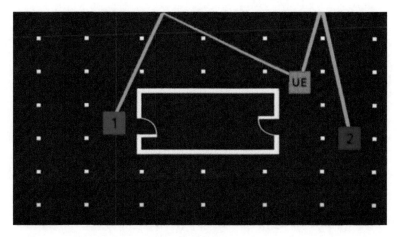

Fig. 18 NLOS communication. Source: Prepared by the author

packets in many different directions, bouncing off buildings and other objects in a precisely coordinated pattern as can be seen in Fig. 18. By choreographing the movements of the packages and the timing of arrival, beamforming allows many users and antennas in one massive MIMO system to exchange much more information at once.

Like massive MIMO adding more antennas to the station's base, more antennas will also be added to the devices, ranging from 4 up to 16 more. This addition of antennas is a key to beam formation, which enables more accurate advanced spatial tracking. This addition of antennas will allow our devices to connect to the best station in your neighborhood to establish a line of visual communication.

17 Device to Device Communication (D2D)

The constant increase in data traffic and the continuous integration of newer, more advanced devices, is the leading cause of congestion or collapse of the network, therefore, telecommunications companies commit to creating and designing new methods to accommodate the high user application demands. Proximity services (ProSe) is a technology in which network devices can communicate with each other, rather than with carrier base stations.

This brings benefits such as increasing bandwidth, reducing consumption of energy and infrastructure costs. Usually, the services of proximity are viewed for the use of security services community, such as a natural catastrophe. D2D communications is one of the key technologies of 5G, it refers to direct communication between two devices, regardless of the network infrastructure. This new method of communication has advantages such as traffic decongestion, decrease battery

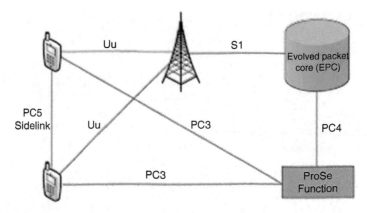

Fig. 19 D2D architecture in logical interfaces. Source: Seet et al. [10]

consumption, increase transmission rates. The 3GPP group has designed a base architecture, which can be appreciated in Fig. 19.

The UE is the user equipment, it wishes to make use of the services of proximity, but previously this must establish a link, with the ProSe functionalities, through the PC3 interface, to obtain security and authorization parameters. When the UE has had an answer of the ProSe function block, it starts in the search process, from another UE with ProSe capability, through the PC5 interface, once these two UEs have established communication, they are linked directly through the "Sideline" physical interface.

In 5G technology, D2D communication can be viewed in two ways, in the first, direct communication between devices is performed without any network entity between these two, in a second way, the UE in charge of retransmitting the signal is between the destination UE and the 5G network. This method is called indirect connection mode, the UE in charge of transmitting the signal to the destination UE, uses access schemes as 5G Radio Access Technology (5G RAT), LTE, WIFI, and networks dedicated. The use of different schemes provides connection facilities depending on the scenario as shown in Fig. 20. In this basic model, different cases can be generated, such as be:

- Case 1: Base station B sends a data signal directly to the requesting node, e.g., (UE1, UE2, UE3... UEn)
- Case 2: The base station B sends data to the main node UE1, and this too, in turn, relays the data to the other nodes (UE2, UE3... on).
- Case 3: In case the nodes require the same information. The base station sends the packets separately to the different nodes, and missing parts are shared through the D2D communication, between nodes.

This technology in the not too distant future will be used for autonomous vehicle communication as well as S2S (Ship-to-ship), which says communication between vessels in the event of a possible accident, to provide more safety and avoid accidents [10].

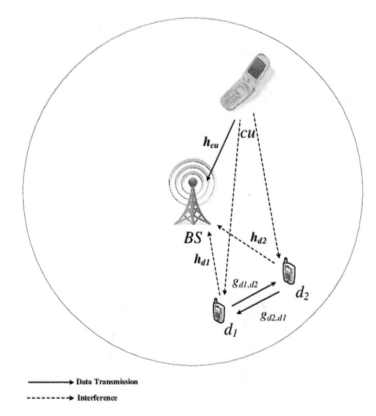

Fig. 20 Simplified D2D model. Source: Seet et al. [10]

18 NOMA Access Technology

Non-Orthogonal Multiple Access is an access technology, which promises to be the most suitable for fifth-generation networks. Unlike NOMA, orthogonal multiple access techniques (OMA), such as TDMA and OFDMA, provide service to a single user in each orthogonal block. Pre-5G networks such as 4G, use OFDMA, this technology has many advantages for eMBB (broadband mobile), however, for IoT or mMTC. This technology is very limiting since it has disadvantages, some are:

- Traffic congestion.
- Interference in multiple access in the uplink.
- Low spectral efficiency
- High energy consumption.
- NOMA is the proposed access technology for 5G, it is expected that this method, will bring benefits like, combination with other waveforms, increased capacity, and spectral efficiency, low latency, traffic decongestion, coexisting connections, less interference, low energy consumption.

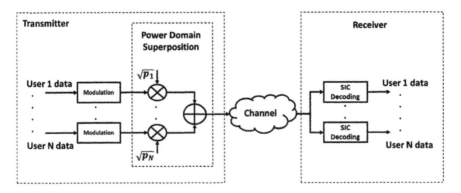

Fig. 21 NOMA diagram

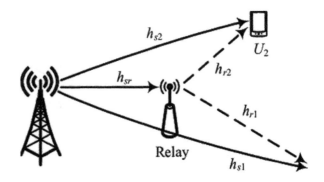

Fig. 22 Cooperative NOMA

Being a technology compatible with other access techniques, it allows make the air interface more flexible and adaptable, for different users with various terminal capabilities. This new waveform promises increased cell capacity, because NOMA, along with other access technologies makes use of frequencies and time. But what makes a robust and efficient system, compared to other access technologies orthogonal multiple (OMA) is that NOMA makes use of an additional domain.

The power domain NOMA is based on the allocation of power to different users, depending on the condition and status of the channel, this allows access to multiple users, it should be noted that this is done in the same slot time, OFDMA subcarrier, or propagation code.

In Fig. 21, a scenario is illustrated where a link of drop for two users (U1, U2), where the base station BS provides service for two users at the same time and on the same subcarrier OFDMA.

As shown in the main Fig. 22, the objective of NOMA is to assign more power to the user with a poorer channel condition. For example, U1 is at the limit of the cell, therefore more power is assigned to it.

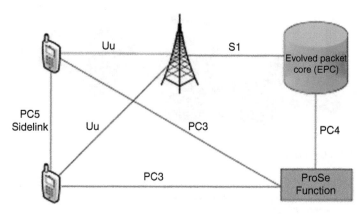

Fig. 23 Full Duplex

19 Full Duplex

Current communications have already implemented full technology duplex, it can use time division duplexing or frequency division duplexing. TDD is a technique in which the downlink as well as the downlink rise is brought on the same frequency by using intervals time-synchronized.

On the other hand, FDD is a method used to establish a full-duplex communication, using two different carrier frequencies, to make the upload and download links. These methods achieve full-duplex communication, however, using different instances of time, or different carriers causes that there are a delay and a loss in bandwidth.

With the integration of Full Duplex for 5G technology, it will be possible to transmit and receive on the same frequency band and at the same time instance is say that both transmitters and receivers will be able to work at the same time and the same frequency, generating greater spectral efficiency.

Considering that this method will use the same time instance, it can be deduced that there will be a reduction in talk time thus improving the flow of communications. Similarly, if one considers that with Full Duplex, the same carrier frequency is used to transmit and receive, consequently the wireless capacity will be doubled, in the layer physics as can be seen in Fig. 23.

For this to be possible, a circuit is needed, which is capable of routing the input signals and output signals in such a way that these do not collide with each other. This is very complex, due to the trend of radio waves to travel both forwards and backward in the same frequency. This principle is known as reciprocity.

As can be seen in Fig. 24, the CMOS circulator chip is a solution to the problem of reciprocity. Allowing signals that will be transmitted and received can work in the same frequency band.

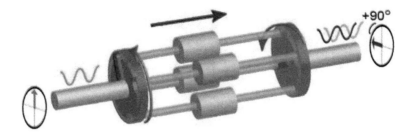

Fig. 24 CMOS Circulator

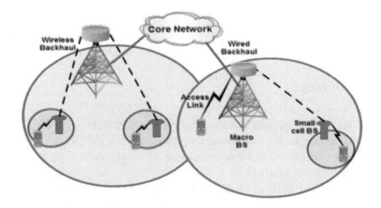

Fig. 25 Massive MIMO, Small cells, MTC

20 5G Architecture

Future fifth-generation networks must be able to support the services and applications that the new demands require, these networks must be endowed with greater flexibility, agility, and scalability, for which 5G network architecture is comprised of various technical enablers keys to meet the needs of users. This architecture is composed of several technologies mentioned above such as D2D,

They will cooperate to carry out the different functionalities and applications of this network. A 5G network is made up of three blocks as can be seen in Fig. 25.

- UE
- NG-RAN
- 5GC

21 NG-RAN

In terms of a network that allows access through radio frequencies, generally refers to base stations that give access to the internet through the transceivers. For example, in GSM, BTS, allowed access to the internet, in 3G the access network was done

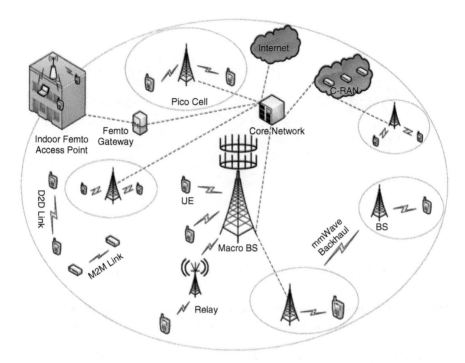

Fig. 26 5G system

through Node B, then in 4G-LTE, Node B evolved to eNodeB, and for 5G, it is has set as gNodeB (New Generation Node B). That is to say that the 5G NG-RAN radio access network to be composed of a set of gNodesB.

The main objective of this new access network is to virtualize it, that is to say, implement a radio access network in the cloud C-RAN.

The C-RAN is composed of two basic elements:

- Control Unit
- Distributed Unit

The grouping of several BBUs is known as a BBU pool of BBUs as can be seen in Fig. 26. This pool acts as the cloud where information from the RRUs will be hosted, or rather, the BBU pool works as a data center, which stores information from the network to meet the operating requirements.

As can be seen in Fig. 27, RRUs act as access points that serve to extend the access network and to give connection to the various devices wireless. Finally, it is very important to mention, the links "Fronthaul," since these links are in charge of intercommunicating the BBUs with RRUs, these links can be fiber optic or wave millimeter, allowing a high transmission speed [11].

In conclusion, one of the new features that will make 5G a new experience is C-RAN (radio access network in the cloud), this new technology allows access to the network through the cloud, that is, there will be virtualization of the network.

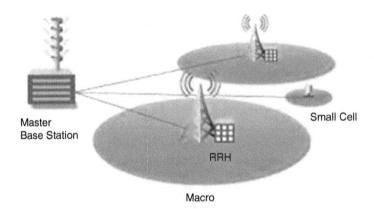

Fig. 27 C-RAN architecture. Source: Zhu et al. [11]

22 Internet of Things, Basic Principles, and Architecture

The internet of things is a disruptive technology capable of giving connectivity to things, that is, objects will be able to access the internet, objects, generate data that will be stored in the cloud. These collected data are processed, analyzed, and subsequently delivered to other objects or devices, in such a way that it generates highly advanced cloud-based service capabilities.

This global connectivity of things is done through the interoperability of various present and future technologies, for these analyzes, and recently conducted studies have considered 5G technology as the ideal platform for its operation, creating and innovating applications and services that facilitate people's daily lives.

Currently, there are already connected objects, these are, for example, laptops, smartphones, smart watches, smart TV, media players, and even smart refrigerators.

In Fig. 28 it can be seen that as time passes, new devices appear, this creates a trend that makes the use of these devices increase. With the continuous integration of new devices connected to the internet, little by little an IoT ecosystem. It is anticipated that by 2020 there will be approximately one trillion objects connected to the internet, according to the Census Bureau of the America, there will be 7.6 billion people by that time, and you can estimate that, for each person, 6.6 objects are assigned which would give as resulted in an estimated 45.6 billion devices.

The planet will be covered by millions of sensors, these will be responsible for collecting information from physical equipment and uploading it to the cloud, where this information will be managed. People will live in an environment with the constant movement of information, that is, there will be a lot of technology that will be transparent to people in their day-to-day lives.

To reflect the location of the internet of things in the vast world of technologies, Fig. 29 reflects a Venn diagram showing three fields that converge with each other resulting in the IoT. In the field of semiconductors, the law established by Gordon Moore, which is a forecast of the year 1965, which states that the number of

Study and Analysis of 5G Enabling Technologies, Their Feasibility and the... 129

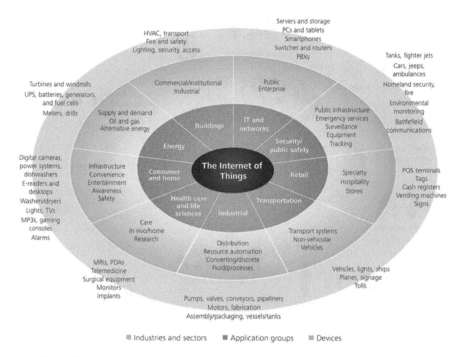

Fig. 28 Proliferation of IoT devices

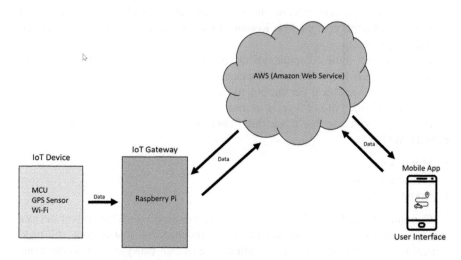

Fig. 29 IoT location. Source: Prepared by the author

transistors that can be placed in integrated circuits is twice as dense, every 2 years. Making computers ever faster and more powerful. This justifies the great leap of the IoT since it is based on moving from devices as a PC to battery-operated equipment such as mobile phones and sensors.

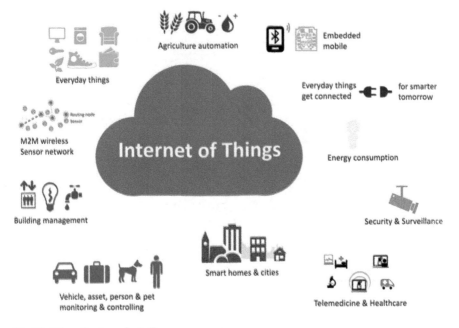

Fig. 30 Disruptive laws for IoT

In Big Data, lies Koomey's law, this is as important as the above since it is related to energy consumption in the IoT. Koomey exposes, the amount of energy required to perform processes.

Computational joule computations double over time every 1.57 years, making computing more energy efficient. IoT scenarios include infinity of sensors, which are not connected to electrical current outlets, which is why it is seen that the devices have batteries with a life of 5 years or more.

In Fig. 30 you can see these three laws that are very essential for an IoT ecosystem [12].

- Moore's Law
- Metcalfe's Law
- Koomey's Law

Currently, the IoT already exists but not in its entirety, since the current devices that have internet connectivity, are not 100% independent, that is, they still need human intervention, the IoT of the future, foresees an ecosystem of autonomous devices with support in 5G technologies, which collects information, without human intervention, in other words, these "things" will be monitoring people's day-to-day life, to make decisions depending on the requirements and existing needs, making them an efficient and useful life [12].

23 Internet of Everything

Throughout the time since Kevin Ashton coined the term "internet of things" in 1999, the internet of things has adopted several concepts and points of view, however, the most basic and accurate, is the one that can describe what their name is, that is, things connected to the internet, without however, lately another IoE technology has been adopted the internet of everything.

The internet of everything is nothing more than a new phase or evolution of the IoT. The internet of the whole is seen as absolute connectivity of not only things, also people, data, and processes make a global network more valuable as shown in Fig. 31.

A global network, connected to everything, aims to offer different types of services, such as critical communications, security, monitoring, medicine, industries, that is to say, that universal connectivity will develop without doubt new applications and opportunities.

24 Industrial Internet of Things

In recent years, large industries have adopted a digital transformation, thanks to the internet revolution. The General Electric company defined the industrial internet as: "The convergence of the global industry with the power of advanced computing,

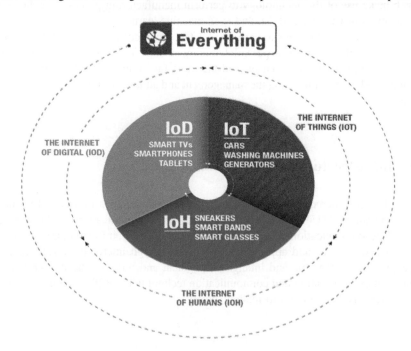

Fig. 31 Entities that make up the Internet of Everything concept

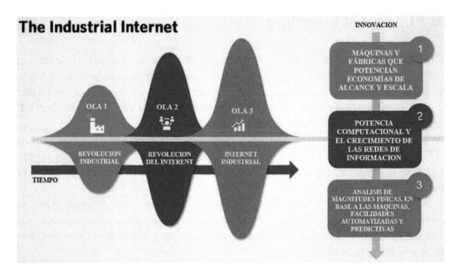

Fig. 32 Growth. Source: Researchgate

analytics, low power detection and new connectivity levels enabled by the Internet [13]."

This under the concept of the three revolutionary waves throughout the time, as can be seen in Fig. 32, bears fruit on what is known such as Industrial Internet of Things (IIoT), which is another of the terms adopted, specifically by industries, which make use of this technology to perform manufacturing processes, collection and analysis of Big Data, M2M, and even machine learning techniques.

It allows large companies to verify and detect any inconvenience or defect in the processes so that there is a potential money-saving. Therefore, a company can ascertain supplies, quality control, and the most favourable, with low power consumption. They are optimizing the management and administration of the company's assets.

25 Infrastructure and IoT Technologies

The IoT is considered one of the most important technologies in terms of technological advances. ITU-T (Industry Standards of Telecommunications of the International Telecommunications Union) has defined the IoT as an infrastructure for access to advanced services and applications achieved through intercommunication. The objective can be tangible and intangible, physical and virtual, whose operation is supported by information and communication technologies (ICT), obtaining a comprehensive advance in the information society.

Fig. 33 IoT dimension

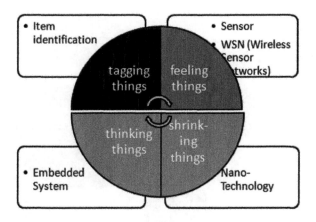

In Fig. 33 you can see a diagram showing three dimensions, it is worth highlighting the OBJECT dimension, that is, there will be communication with anything or object, regardless of the PLACE and INSTANT.

26 Convergence of the Physical and Virtual World

In the internet of things, two worlds reside, of which are composed of entities or objects, these objects, as mentioned they can be physical and virtual, therefore, in the IoT, there is a convergence between the physical and virtual world, in the physical world are the tangible objects, for example, tablets, laptops, webcams, Smart TV, and everything type of tangible object. On the other hand, there are the intangible objects that reside in the world virtual, from the IoT point of view, these objects are considered as the information, such as files, software, online applications, etc. By unifying these two worlds, this information will be the path to the different applications and services.

In Fig. 34 you can see how the IoT works. A device (Physical), can be associated with several virtual objects, that is, that there is a one-to-many relationship, however, a virtual object, not necessarily has to be associated with a physical object.

These physical objects are equipment that has capabilities communication technologies, they are also in charge of carrying out the detection and acquisition functions, storage, processing, and information actuation. In other words, objects are capable of executing the actions according to the data received by the networks of communication.

These devices establish a link with the world of information, through communication networks, in which there are gateways or gates in charge of managing the information, and its subsequent translation of the information, to the necessary protocols of the communication networks.

However, two devices can communicate with each other, without the need for a gateway, also through communication M2M, D2D, mMTC, two devices can communicate with each other without the need to access communication networks.

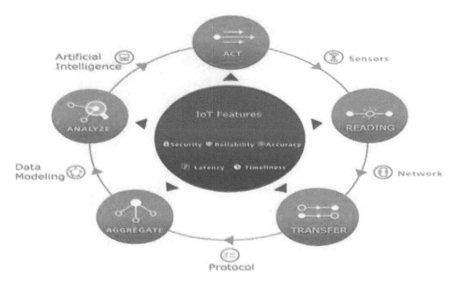

Fig. 34 IoT operation

It should be noted that this explanation is based on the physical world. However, in the virtual or information world, there may also be mutual communication between virtual objects so that the data is communicated with the physical world.

Communications networks are responsible for transporting information collection of the devices, towards the applications. These blocks of information, which is collected, are instructions from the applications to the devices, in such a way that they are activated, making decisions. Communications networks have transfer capabilities the data acquired, making them efficient and reliable. The IoT can be developed in traditional networks, such as TCP/IP, or evolutionary networks such as the famous NGN (Next-Gen-Networks).

27 General Structure of the IoT

The IoT is composed of three very important pillars, which are:

- Sensors
- Network connectivity
- Data storage (Big Data)

In Fig. 35 you can see an IoT structure, this is composed mainly of "things," when speaking of these refers to the sensors integrated into them such as humidity sensors, temperature, presence, motion, health sensors, among many more, depending on the application or service that is being used. These are in charge of detecting and

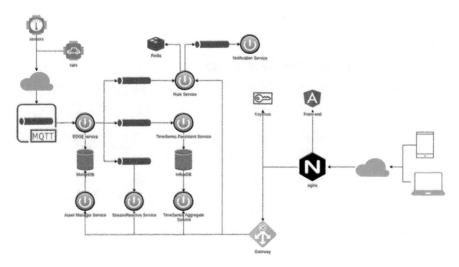

Fig. 35 IoT structure

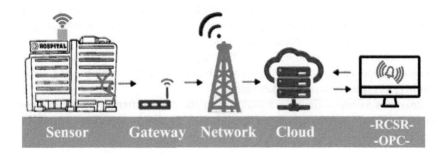

Fig. 36 Ideal IoT Scheme

measuring some physical magnitude, generating an analysis of certain information, digitizing it, and sending it to the cloud.

However, currently with the aforementioned networks 100% ideal operation of the IoT is not achieved, so the networks of fifth-generation 5G will serve as support for these networks to meet the key capabilities for the proper functioning of IoT, that is why a 5G network is considered a key enabler for the continued development of the IoT.

In Fig. 36 you can see the ideal scheme of the IoT where the sensors send the information or data to a local processing unit and later to a local storage unit.

The problem today is that the sensors send the information collected directly from the network is to say that a lot of information will go to the cloud which is irrelevant for it since it does not process them and does not perform decision-making. On the contrary, if the information passes to the processing unit, the data sent to the unit can be analyzed and collect key information, which will be sent to the cloud, that is, this information is correct for carrying out decision-making.

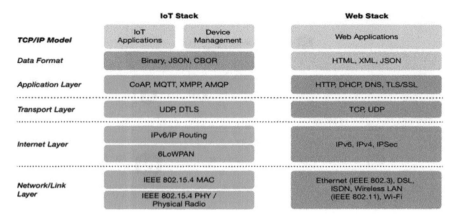

Fig. 37 IoT Protocol stack. Source: Gerber [14]

28 IoT Protocols

The protocols adopted by the IoT fit within the layers of the TCP/IP reference model, shown in Fig. 37. The high demand for new applications and services, emerging new technologies, and interconnection methods, for which operators and service providers, are forced to work with a 100% adaptable interoperability, creating and innovating new network technologies, which are gradually adapting to the ecosystem IoT, targeting different market niches.

Providers create different network protocols, each differing by its structure, topology, and operation, depending on a condition, specific service, or application [14].

29 IoT Networks

The idea of the internet of things dates back to 1983, when the idea of creating a network for smart devices, took initiative in the creation of automated inventories. However, it was in 1999, when the IoT emerged as a future technology.

On the other hand, there is the IIoT, which focuses on the integration of two technologies, these are the operational technology (OT) and the technology of information (IT), through this convergence a data can be obtained analytics, through an intelligent sensor network, made up of an infinity of intelligent machines, to improve b2b services (business-to-business), targeting different market sectors that go from manufacturing to utilities.

Although each of these has different applications, both have the requirements to cover the general communication requirements as it is, adaptation to the IP ecosystem. Therefore, the networks of communications must be resistant to packet loss; be safe and resistant to damage, and, more generally, achieve the balance desired between capital expenditures/operating expenses (CAPEX/OPEX) and system/service availability.

Cellular networks, through many studies and analysis, have proven to be the most promising technology for the development of the internet of things. The 3G, 4G, and 3GPP LTE networks have served as support to certain IoT applications. However, as already mentioned, MTC is the core technology of the IoT, over which 4G and even 4.5G are not capable of supporting this technology.

30 Edge Computing

When various information and things are connected to networks, this is referred to as the internet of things, and the massive and incomplete data generated by IoT must be processed and responded to in a very short period of time. Today the cloud has become an indispensable part of that process; however, the cloud that has been centrally deployed at scale global needs to process an enormous amount of data and as increases the physical distance between the user and the cloud, the transmission latency, that is to say, the response time increases, furthermore, the processing speed in this environment is highly dependent on the user device's performance.

The solution to these issues is edge computing, which works by allowing a small server located between the cloud and the user to perform application processing, and crucially, it is physically closer to the user., this enables the necessary information to be downloaded from the cloud or the user's device at a location closer to the user for processing, thereby accelerating applications that require low latency response.

As shown in Fig. 38 the edge computing layer represents the infrastructure located closer to endpoints and includes a collection of MDCs (Micro Data Centers) geographically distributed. The architecture primarily provides a solution for applications that have a higher sensitivity to latency and require processing and storage.

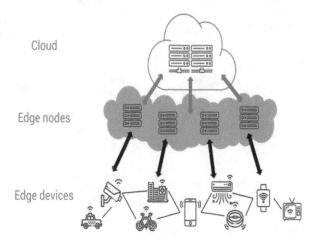

Fig. 38 Edge Computing Model

Here, small-scale MDCs that represent a very important role in the information collected and distributed pose new challenges in terms of efficient management and distribution of resources to achieve efficient resource allocation.

31 5G Security

The next step in the evolution of communication mobile and 5G will also be a fundamental enabler for the network society, so it will have to provide capabilities far beyond what is currently available, with new users and new use cases, thousands of millions of devices and new applications to connect society in general, this also means that new approaches to defining safety for mobile will be required.

As previously stated, the future 5G network would define unique throughput and latency specifications to meet the needs of various industries such as smart transportation, smart grids, and so on which can be seen in Fig. 39. There will be machines and sensors connected to the 5G network, where new models of network

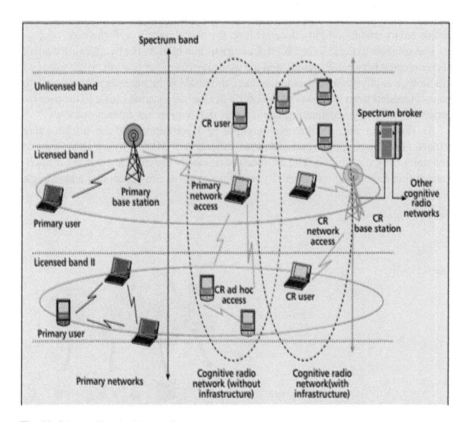

Fig. 39 Licensed band xG network

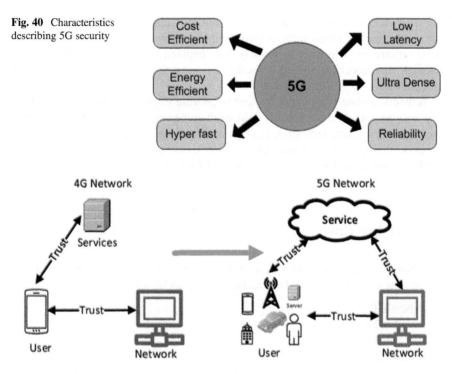

Fig. 40 Characteristics describing 5G security

Fig. 41 4G vs 5G Trust Model. Source: Huawey [15]

and service infrastructure will be provided, which will have an impact on the definition of 5G security.. When GSM systems were first introduced 25 years ago, protection was designed to protect connectivity and voice, with later generations including data. The key goal was to gain user confidence in terms of privacy while also safeguarding the environment in terms of security. This has performed well, but 5G would necessitate additional criteria.

In Fig. 40 we can see the characteristics that entail when the development of a new security platform for the 5G network. Authentication only by the service provider, on the other hand, networks can rely on the capabilities of verified authentication of industries and devices exempt from radio network access authentication, which can help networks to reduce operating costs.

Network and service provider authentication a legacy model can be used for some of the facilities. Networks are in charge of network access, while service providers are in charge of service access [15].

As can be seen in Fig. 41, With 5G the new service delivery models; the use of the Cloud virtualization and network segmentation emphasizes the need for secure software, and have multiple applications running on top of it same hardware, they put requirements on virtualization with strong insulation properties 5G will play a prominent role as critical infrastructure, and attacks against the 5G network could have serious consequences for society, this evolved threat environment requires strength and security, and also shows the need for measurable security and compliance.

Network Slicing is an important tool to provide isolated subnets, optimized for applications with different needs and different user groups, but could also be used to meet specific requirements, from a safety and security perspective. Finally, related to network slicing, virtualization, configuration dynamic and software-defined networks require an architecture of dynamic and flexible security.

32 IoT Use Cases

As can be seen in Fig. 42, there are several fields on which the IoT can work, offering different types of services and applications, these are as follows:

- Public IoT
- IoT in industry
- Personal IoT
- IoT for the home (Domestics)

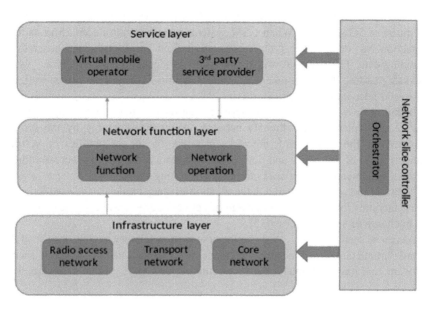

Fig. 42 Network Slicing

33 Public IoT

It should be noted that the deployment of these applications can be performed through the implementation of NB-IoT or LTE-M, that is, the LPWAN technology plays a great role in the development of these services.

34 IoT in Industry

IoT in Industry mainly offers wide-area applications of low power that help improve overall business efficiency and industrial. Here are some examples

35 IoT at Home

Conventionally, smart home applications are implemented in short-range technologies such as Z-Wave, Zigbee, but they are you need a home gateway. In the case, where the device is.

Integrated with an NB-IOT chipset, the benefits are amazing. For example, administration becomes more efficient through improvements in big data analysis. The IoT device is mainly composed of LPWA applications that aim to provide intelligence for the user through sensors and devices found in the local area.

36 Conclusions and Suggestions

5G technologies and the internet of things are among the main elements that will shape the future of the internet in the next years. In this document, a detailed analysis of the potential of the 5G technologies defined by 3GPP for the IoT, placing them in the context of the current connectivity landscape for IoT. Unlike the earlier cellular technologies, which were essentially designed for broadband, the requirements that future 5G networks will have to meet, and particularly those of MTC, make 5G communications particularly suitable for IoT applications. By offering a minor cost, lower power consumption, and support for a large number of devices, 5G is ready to enable the IoT. 5G can, and will, be a cornerstone in realizing the vision of the Connected Society, in which everything that can benefit from a connection will be connected. But instead of tackling 5G security by trying to implement all imaginable security mechanisms, you need a systematic and analytical multi-stakeholder approach, focused on a new model of trust for 5G networks. This will offer an evolved and flexible 5G security architecture that can provide a trusted platform for this vision.

37 Recommendations

Because the 5G network is currently in phase 1b established by the 3GPPP group and in addition to being a future network, recommends an exhaustive and continuous study in the next requirements and standards that come to light, to solve the possible and challenging scenarios that can be achieved with the development and implementation of the IoT within an ecosystem with 5G technologies, to a vision of a rigid and seamless platform where absolutely everything is connected.

And as a recommendation to carry out an implementation and integration of these technologies in Ecuador, the different standards proposed by both the ITU and the 3GPPP group, to meet the primary and essential requirements which are the key to being able to hyper connected the world and provide a seamless experience.

References

1. Virgen, L., & Sierra, M. (2013). *Alexander Graham bell patents the phone | University of Guadalajara*. Retrieved on June 10, 2018, from http://www.udg.mx/es/efemerides/14-de-enero-de-1876-
2. Vega, M. C., Vivas, P. O., Ríos, C.M., & González, C. 5G Security: Retrieved from Forward Thinking Huawei White Paper. Recovered From https://www.huawei.com/minisite/5g/img/5G_Security_Whitepaper_en.pdf
3. Adell Hernani, JA, & Telefónica. (2002). *Telecommunications again generation*, Madrid.
4. Dahlman, E., Parkvall, S., & Skold, J. (2013). *4G: LTE/LTE-advanced for Mobile broadband*. Elsevier Science. Recovered from https://books.google.com.ec/books?id=AbkPAAAAQBAJ
5. Seurre, E., Savelli, P., & Pietri, P.-J. (2003). *GPRS for Mobile Internet*. Artech House.
6. Brand, A., & Aghvami, H. (2002). *Multiple access protocols for mobile communications: GPRS, UMTS, and beyond*. Wiley.
7. Triggs, R. (2017, June 3). What is LTE advanced? - Android authority. Retrieved on March 8, 2021, from https://www.androidauthority.com/lte- advanced-176714/
8. Mumtaz, S., Rodriguez, J., & Dai, L. (2016). *mmWave massive MIMO: A paradigm for 5G*. Elsevier Science. Recovered from https://books.google.com.ec/books?id=rg1YCwAAQBAJ
9. Marzetta, T. L., Larsson, E. G., Yang, H., & Ngo, H. Q. (2016). *Fundamentals of massive MIMO*. Cambridge University Press.
10. Seet, B. C., Hasan, S. F., & Chong, P. H. J. (2018). *Recent advances in cellular D2D communications*. MDPI AG.
11. Zhu, Z., Gupta, P., Wang, Q., Kalyanaraman, S., Lin, Y., & Franke, H. (2011). *Virtual base station pool: Towards a wireless network cloud for radio access networks* (p. 1). ACM Press.

12. Geng, H. (2017). *Internet of things and data analytics handbook*. Wiley. Retrieved from https://books.google.com.ec/books?id=V3mYDQAAQBAJ
13. Evans, PC, & Annunziata, M. (2012). *Industrial internet: Pushing the boundaries of minds and machines*.
14. Gerber, A. (2017). *Connecting all the things in the internet of things*.
15. Huawey. (2015). *5G security: Forward thinking huawei white paper*.

Automated Methods for the Detection of Green Land in Satellite Images

Raju Pal, Subash Yadav, Aarti, Pushpendra Kumar Rajput, and Anand Nayyar

1 Introduction

Semantic segmentation, sometimes called object classification when dealing with the field of remote sensing (RS) is the method through which pixel classification of an image with its corresponding label is performed. It is an essential task in the field of object recognition and has many applications [1]. With the speedy improvement in the field of RS technology, high-resolution remote sensing satellites, namely Quickbird, SPOT-5, WorldView, and IKONOS have provided us a plethora of information needed for recognizing the ground objects, in a way that we never could before when using low-resolution remote sensing imagery. Due to the advent of these new technologies, many object classification problems in which objects are difficult to be recognized due to limited amount of information can now be easily detected [2–9]. Semantic segmentation is now being widely used in various computer vision application areas particularly through the use of shallow features by people who have experience and expertise in the particular domain [10]. The problem with this methodology was the fact that if the conditions do not remain the same and change even slightly, the pre-existing framework which has been developed and which works well the given task may not work in another task and

R. Pal · S. Yadav
Jaypee Institute of Information Technology, Noida, India

Aarti (✉)
Lovely Professional University, Phagwara, India

P. K. Rajput
University of Petroleum and Energy Studies, Dehradun, India

A. Nayyar
Graduate School, Faculty of Information Technology, Duy Tan University, Da Nang, Viet Nam
e-mail: anandnayyar@duytan.edu.vn

fail which would lead to the whole feature extractor being trashed and would have to then be rewritten from scratch. This will be an expensive and time-consuming task. These disadvantages led the researchers in the field towards the pursuit of a more effective and robust approaches [11].

Moreover, semantic segmentation can be used for automated object labeling, and its ability to interpret environments comprised of many different object classes [12]. This means it is now being used in a plethora of state-of-the-art technologies such as facial segmentation to help computer vision systems recognize and make predictions about a person's expressions, their age, their mood, etc.; it can be used for autonomous driving, by identifying the obstacles in the road and the regions like roads and lanes; and satellite image processing along with land usage analysis by identifying and labeling the different regions for planning and management purposes [13].

Semantic segmentation is also being increasingly used in the RS field, due to its ability to deal with large datasets and its high performance. It is the mechanism through which information about any objects or any phenomenon is gathered without ever making any physical contact with the location containing the said objects. It is very useful to gather data from places which before were thought to be uninhabitable and places where it was traditionally unfeasible to go in order to collect data. Due to the onset of satellites which are armed with high-resolution cameras and telescopes, we are now able to look in places that we never could before, like the arctic or the polar ice caps [14]. These technologies provide us with various different types of data, one of which is satellite images. These satellite images which can now be taken from the comfort of the researcher's office have now opened a Pandora's box which provides us with such high-quality images which can help us in observing many different phenomena taking place all over the world, like classifying vegetation [15], land mapping [16], and for tracking soil erosion all by using image labeling via semantic segmentation. Now equipped with various machine learning methods at hand, and by combining it with semantic segmentation, we have at our disposal a method through which we can label and monitor green land all over the world and invent new plans and approaches for the betterment of the world. This model can also be helpful in battling global warming.

Global warming stands right now as one of the biggest challenges our humanity has faced till now, and it is caused due to increase of greenhouse gases. Historically, nature had a remedy for the removal of these greenhouse gases, i.e., trees, but due to many different reasons, we have cut those down for our own personal benefits. This will lead to increase in global warming and it is important to gauge the impact of the drastic weather changes in the amount of landmass which is covered by trees. By figuring out which places are devoid of forestation, we can then plan how to repopulate that barren land with trees in order to maintain the homeostasis in the ecosystem. Therefore, it is required to identify green areas in particular region from satellite images so that planning and managerial decisions can be made for increasing these areas. The main objective of this work is to detect the green areas present in the given satellite image using semantic segmentation techniques. Our model tries to calculate the area of the landmass which is covered by trees.

In literature, convolutional neural networks (CNNs) have been extensively utilized for several image recognition problem in order to obtain tremendous results [17–19]. Recently, U-Net architecture [20] has generally provided better results than other traditional models in the semantic segmentation field. To further improve its performance for satellite image segmentation, we have combined a U-Net with a residual neural network (ResNet) [21], which works in conjunction with the U-Net to provide better results by skipping one or more layers [22]. When combined with the data in the current layer turn, the data from the previous layer helps create Residual Blocks. This hybrid model helps in limiting data loss as well as in the creation of a deeper overall structure. In residual blocks, the output of the previous layer is fed as input to the current layer, which helps to minimize the data loss of the previous layer, as the resultant output is a function of the current and all the previous layers. This would lead to better accuracy and less data loss.

Therefore, the main objectives of this chapter have two folds:

1. A new hybridized U-NET model is introduced with ResNet.
2. The proposed model, namely skip-U-NET, is used for semantic segmentation of green lands in satellite images.
3. The experimental analysis of the Skip-U-NET model will be tested and analyzed on DSTL satellite image dataset.
4. For better comparative analysis, the performance of the proposed model is evaluated against other state-of-the-art methods.

The chapter organization is given as follows: Literature survey has been presented in Sect. 2. The characterization of the standard U-NET method is provided in Sect. 3. Section 4 presented the proposed U-NET method. The experimental results of the proposed method are provided in Sect. 5 followed by the conclusion of the chapter.

2 Literature Survey

In the literature there are mainly three types of deep-learning methods used, namely simple classification-based method, CNN based method, and fully convolution network (FCN) based method. In early of 1989, the main focus of the researchers was on solving pixel level classification using some machine leaning algorithm for the remote sensing images [23]. The main approaches behind the early research, the features chosen were the value of different spectral band. Some of the famous handcrafted features which comes from the textural and local spectral in form of color histograms, HOG, and SIFT had taken into consideration [24]. The famous method for the image classification is the Bayes' classifier [25]. Some of the popular classifier as SYMs [26], random forests [27], and different types of boosting [28] have become the backbone for the application of machine learning to high-resolution imagery for learning this non-linear decision boundary. The classifier did not have enough information by the approach of use of hand-crafted features and value of multiple bands in the modification of resolution. Both precision and recall get improved significantly with supervised and unsupervised learning methods, namely restricted Boltzmann machine [29], sparse coding, and CNN.

The classification is the primary use of CNN and each output of the classification contained the different label of the class. CNN based framework mainly used for implementation of semantic segmentation is proposed by [30]. For the training or the prediction of label of the pixel, the input for the CNN model is the image patch extracted near each pixel. Extraction of the patches around all pixel is done by sliding window in the input image. The either way of using modern GPUs implemented CNNS for the learning of image feature of high discrimination. The difference between the patch based and simple classification algorithm is of accuracy with patch based on higher accuracy [31]. There are some disadvantages too of this method as higher processing time, large memory requirement and there is also boundary for receptive field. Therefore, Shelhamer et al. [31] proposed an FCN based framework for the segmentation of the image. In this all the fully connected layers in CNN are removed and the recovery of each pixel based on specific classification from abstract features is done. FCN has input and output as images. That is why it is an end-to-end model. The pixel positioning and pixel classification problem is solved using high-resolution shallow layers and low-resolution deep layers. FCN is considered a research hotspot in semantic segmentation. Compared with the traditional method, it is more effective and can fit any size of input and output of segmentation. Increases in the benchmark datasets draw the researcher's attention to the ML system's better development for semantic segmentation [32]. Recently in RS, many researchers focused on the FCN based semantic segmentation. Ivanovsky et al. [33], presented developed CNN architecture that belongs to these neural networks which aimed at the different object classification and detection of fast and high quality. In the model, CNN was the part of deep machine learning algorithm which is very helpful to solve the recent computer vision problem in a particular domain of satellite image segmentation. There are mainly two types of popular models, i.e., U-Net and LinkNet for semantic segmentation. U-Net has mainly two parts, namely decoder (right) and encoder (left). Similarly, LinkNet also has encoder decoder configuration.

Further, Ghosh et al. [34] trained a deep machine learning architecture consisted of stacked U-Net to perform the classification of remote satellite images. Here, four blocks of stacked U-Net are used which contain 2,7,7, and 1, respectively. The input and output have size as 512×512 and 32×32. Bilinear interpolation is used to rescale the output to original size. The loss function used is the multi-cross entropy. For training, Adam optimizer was used with the starting learning rate of 2.0e4. Moreover, Iglovikov et al. [35] discussed about the semantic segmentation of satellite images using U-Net. They especially work on the Kaggle competition dataset based on satellite imagery. The dataset has two categories, namely public and private part. At the first stage, the public part is used in model evaluation while final model got tested on private part for model evaluation. The dataset contains 57 images which is particularly divided as train (25 images) and test (32 images) sets. The class of the images classified on the basis of band width high, medium, and low. Here each pixel has got a class after assignment. Also target classes got imbalanced, i.e., many other classes are included with each pixel. Therefore, separate model for class is good over single model for prediction of all classes. This research is especially based on the infrared images and the algorithm works on the color to identify the classes. That is why the best result obtained in the case of water and vegetarian classes based on the reflectance index. For the problem of overlapping pitches which is previously solved by cropping on the edge but in this chapter a better way was used. The cropping layer is added to the output layer which simultaneously solves two problems, i.e., losses occurring on edge artifact are not held through the network and the automatically cropping has been done of the edges of prediction which result in the decrease in the computation time. This research has trained the model for six classes, i.e., defined earlier. This research mainly includes FCN to satellite with analyzation of boundary effect and joint training objective and the reflectance in the dices. As earlier mentioned, that the result of this has been compared on two sets, i.e., public and private set which yield a better result.

Again in [36], Wu et al. mainly focus to identify the refugees' camp to provide humanitarian support all over the globe. They chose the dataset from the Kaggle DSTL competition and the crowdedAI mapping challenge. Hu et al. [37] also discussed three methods, namely R-CNN, U-Net, and morphological operation. They used NVDIA GeForce 1080 Ti GPU for training. There are losses occurring every four epochs of training in the R-CNN. In U-Net there is improvement in IOU but takes longer time to run. The best result was obtained by the U-Net model with dense connection. The future work suggested by this paper is to use VGG-16 and ResNet model which surely perform better. Moreover, Culberg et al. [38] proposed a method to automate the analysis of key features in online satellite images by classifying the satellite images pixel-by-pixel into one of the eight predetermined classes, such as buildings, roads and crops, etc. To predict the feature classes, they trained logistic regression models as well as Support Vector Machines (SVM's) and applied it upon the multi spectral satellite images. While doing this, they discovered that on the basis of the Jaccard Score, SVM performed worse than the logistic regression model, with both models giving better results on the crops, waterways, and building classes than the other classes. Kodruch et al. [20] developed a model called TreeNet, which is used for classifying the trees that are in satellite images on a pixel-by-pixel level. The model was built on top of U-Net architecture, which provides the ability to precisely localize elements in the field of bio-medics. Their main aim was the identification of inherent challenges in the field of semantic segmentation or the pixel level classification. They developed a mask which when fed satellite images would output black and white images with the white portion signifying the presence of trees in the image. When developing the model, two methods were presented, either they could use the DeepLab V3 which is a state-of-the-art image segmentation method, or use a U-Net based architecture used in biomedical image segmentation. Due to U-Net's ability to consume less parameters and its relatively faster speed while training and its less power consumption, it was preferred over DeepLab and other DCNN models. Khryashchev et al. [39] made three modifications of CNN architecture to segment satellite images, namely U-Net,

SegNet, and LinkNet and compare these modifications' results. The results show that by using complicated CNNs, we are able to increase the quality of segmentation of satellite images. U-Net provided 94.66% accuracy followed by LinkNet with 94.53% and SegNet with 93.59%. The Dice Similarity Coefficient (DSC) was 0.45 for SegNet, 0.68 for LinkNet and 0.75 for U-Net. Furthermore, Avenash et al. [40] proposed a way to use U-HardNet which is a form of CNN by using Hard-Swish for segmenting the satellite imagers. They also try to achieve binary entropy loss minimization by using HIS transformed images. In their suggested architecture, they replaced the fully connected layers with convolution layers in order to output spatial maps instead of classification scores. This new idea is termed as U-HardNet which consists of a new activation function termed as Hard-Swish. Faster training is achieved due to the reduction of parameters by replacing the fully connected layers with convolution layers. It also allows for segmentation of arbitrary sized images an end-to-end manner of CNN, which in turn allows precise localization by letting the model combine several low-level feature maps with higher-level feature maps. In the up-sampling part of the U-HardNet, due to using a large number of feature channels, we are allowed the usage of context information in higher resolution layers. This in turn helps keep the computational cost low when doing semantic segmentation by having a smaller number of parameters, as there are no fully connected layers. The Jaccard Index was used for evaluation purposes. On the DSTL dataset, it was observed that Hard-Swish outperformed some other traditional functions. Best evaluation accuracy went up to 97.75% with minimum loss as 0.08%. It received a 0.265% error percentage on MNIST dataset and a 96.1% accuracy on the CIFAR10 dataset, which when observed against other activation functions, proved to be the superior one. Similarly, Osin et al. [41] developed an algorithm that would be used for pattern recognition which would be able to consolidate data from visible and infrared spectral ranges while at the same time allowing users near real-time performance on embedded systems with infrared and visible sensors. The main reason for doing this is because of the high computational cost that is unfeasible when working with low-powered embedded systems which lack any dedicated GPU supporting the task, while at the same time creating an algorithm that would be able to provide acceptable recognition when working in unfavorable conditions, such as poor lighting conditions like night-time for instance. Adam optimizer is used as the algorithm of optimization of numerical having a learning rate of 1e-3 had been chosen. The loss function used is the binary cross entropy. After running on a network batch of 18 samples, the weight of the model is updated. Here the training consists of total 96 epochs for completion. Planet database satellite images are used for numerical experiment. Building belonging pixels are white in color where all information of the images is stored in JSON file and masked black-n-white. The main classes used in this are building and not building. Results of the following have accuracy as 96.31% in U-Net and 95.85% in LinkNet. Worst result is obtained by LinkNet and best result is by U-Net.

Based on the extensive literature review discussed above, it can be observed that U-Net performs better for image segmentation in different application areas as compared to other deep-learning based counterpart. The accuracy achieved by the

U-Net architecture is tremendous especially in case of satellite imaginary. However, there is still a scope of improvement in the existing U-Net architecture. Therefore, in this chapter a hybridized U-Net model is presented which uses the residual blocks concepts of ResNet model to provide better results by skipping one or more layers. The data from the previous layer when combined with the data in the current layer turn helps in creating residual blocks, this hybrid model helps in limiting data loss as well as in the creation of a deeper overall structure.

3 The U-NET Method

U-Net is a type of CNN which was developed earlier by Computer Science Department of The University of Freiburg, Germany [20] for the biomedical image segmentation. Later the architecture extended for the working with fewer images after modification and result in a very precise segmentation. The main idea behind U-Net is to provide a successive convolutional layer to enhance the resolution of the images. This extra layer also precisely assembles the output after leaning [20].

U-Net consists of many feature channels in up-sampling part which is responsible for the context information propagation to higher resolution layer. This network has expensive path and contracting path which is more or less symmetric and responsible for the U shape architecture of this network [40]. The contracting path has repeated convolution application where each followed by a max pooling operation and a rectified linear unit (ReLU). At the time of contraction, there is loss in decrement in spatial information while increment in feature information.

U-Net consists of encoder and decoder. By the process of pooling layer, spatial dimension is reduced by encoder and by the process of up-sampling layer, detail and spatial dimension of the object is restored by decoder. The modified architecture of U-Net is shown in fig. As defined earlier the contraction path and extension path, there are many similarities between the contraction path and traditional CNN in which doubling of the feature channel by max pool happens with a 3×3 convolutional kernel after convolutional layers. For up-sampling operator, extension path and corresponding contraction path feature map merged. The up-sampling process is an up-convolution consisting of 2×2 convolution layer. For makeup of the lack of information misplaced by the maxpool layer, a two-convolution layer along with 3×3 convolution kernel is used. The process of normalization is done during the downscaling path and then following by dropout operation. Cross entropy function is used here. Activation function is ReLU (Rectified Linear Unit) in all convolution layers. Optimizer is Adam and the evaluation matrix is the average Jaccard index.

U-Net has many applications in image segmentation of like pixel wise regression [3], liver image segmentation [41], brain image segmentation [2], sparse volumetric segmentation, satellite image segmentation, and many more.

4 Skip-U-NET Model

For semantic segmentation, the U-NET model is currently the most used and easily sizable or scalable architecture. This network shows more accurate segmentation with input of very few training images. The U-NET model consists of a contraction path (or left part) and extraction path (or right part) with a symmetrical structure.

4.1 System Model

In our proposed model, we have added an extra layer in the typical U-NET model and hybrid model with the concept of Residual Block. The contraction path is the same as a specific convolution network. The number of feature channels gets double after two convolutional layers with a 3×3 convolution kernel in the max pool layer. Up-convolution of 2×2 convolution kernel is used in the process of up-sampling. We have used a dropout of 0.5 in some of the layers of our model. In proposed model, ReLU is used in every convolution layer; the Adam (Adaptive Moment Estimation) optimizer with learning rate 1e-4 is used with a binary cross entropy loss function. The architecture of our model is shown in Fig. 1. Also, the Jaccard coefficient is used as an evaluation metric.

The most common optimizer in CNN is Stochastic Gradient Descent. Still, this method's disadvantage is that it cannot find an appropriate learning rate and converges to the local optimum very quickly. In the proposed model, Adam is used to calculating the adaptive learning rate of each parameter. The Adam optimizer adjusts each parameter's learning rate according to the first-moment estimation and the second-moment estimation of its gradient function of the loss function. Adam optimizer's advantages are that it runs fast, slowly converges, fluctuations in the loss function, and can correct the issues of the disappearance of the learning rate. The first-moment deviation is given as:

$$\widehat{m}_t = \frac{m_t}{1 - \beta_1^t} \qquad (1(1))$$

And the second moment is given as:

$$\widehat{v}_t = \frac{v_t}{1 - \beta_2^t} \qquad (2(2))$$

The Adam optimizer is calculated as:

$$\theta_{t+1} = \theta_t - \frac{\eta}{\sqrt{\widehat{v}_t} + \varepsilon} \widehat{m}_t \qquad (3(3))$$

where n is step and e is a little constant of default value 1e-8.

Fig. 1 Control flow diagram of proposed method

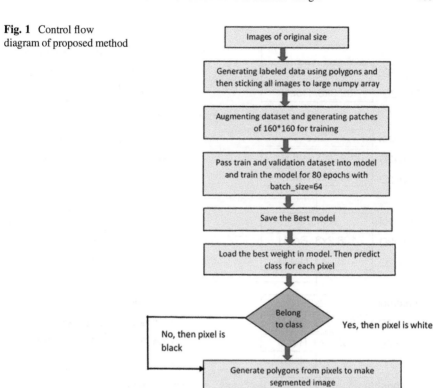

4.2 Architecture and Working

U-Net consists of an encoder and decoder. By the process of pooling layer, the spatial dimension is reduced by the encoder. By the process of up-sampling layer, the details and spatial dimensions of the object are restored by the decoder. The modified architecture of U-Net is shown in Fig. 2. As defined earlier in the contraction path and extension path, there are many similarities between the contraction path and traditional CNN. The doubling of the feature channel by max pool happens with a 3×3 convolutional kernel after convolutional layers. For the up-sampling operator, the extension path and corresponding contraction path feature map merged. The up-sampling process is an up-convolution consisting of a 2×2 convolution layer. For the makeup of the max pool layer's lack of information, a two-convolution layer along with a 3×3 convolution kernel is used. The process of normalization is done during the down sampling path and then following by the dropout operation. Cross entropy function is used here. The activation function is ReLU (Rectified Linear Unit) in all convolution layers.

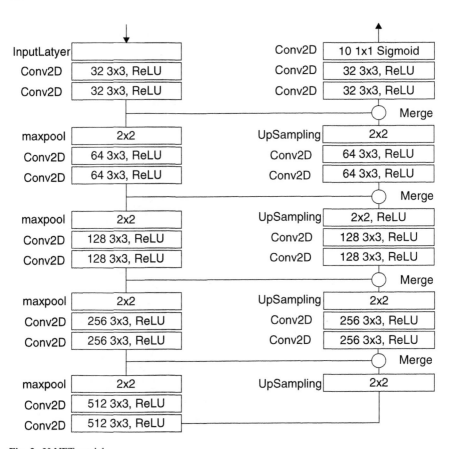

Fig. 2 U-NET model

In our model architecture, we combined the concept of ResNet in U-NET. The main idea of ResNet is to skip connection of one or more layer. We have used concept of Residual block in our architecture by adding the input from previous layer to current output and then pass it as an input to the next layer as shown in Fig. 3.

In the designing of our model, our main focus is on identifying pixel to detect green area on the basis of classes as tree and crops. Input for the proposed model is 8 band images of the M band wavelength. The polygon structure given in dataset is used in generating labeled data and then all images are stored in a large NumPy array. Now by augmenting dataset, patches of 160×160 are generated for training. Now passing these train and validation dataset into our model and training the model for 80 epochs, the best model is saved automatically. Now the prediction of classes can be done for each pixel by loading the best weight. If the pixel belongs to that particular class, then that pixel is white for the masked image otherwise black.

Fig. 3 Skip connection

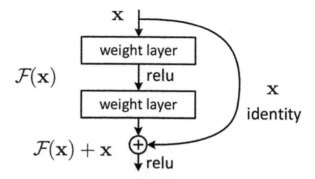

5 Experimentation and Analysis

5.1 Experimentation Platform

We have implemented our model in Keras using TensorFlow backend. We executed our model in Kaggle platform with 16 GB RAM, 5 GB Hard Disk, 16 GB GPU. The implementation is done in python. The requirements of all python packages are Tifffile, TensorFlow, OpenCV, Shapely, Matplotlib, Os, Random, SkLearn, Pandas, and NumPy.

5.2 DSTL Dataset

DSTL Satellite Imagery Feature Detection database is used to train the model in this chapter which is published by DSTL (Defense Science and Technology Laboratory) containing labeled sample data of 10 type. There are 25 1 km × 1 km images available in both 3-band and 16-band format. Satellite WorldView-3 is the source of the dataset which cover an area of 25 kilometers containing 25 images. 3 band images used in this are traditionally a natural RGB color images, whereas 16 band have a wider wavelength consisting of wider wavelength. Table 1 is the description of band data.

The description of the dataset can be as that it contains many different objects like vehicle, road, farms, building, waterways, tree, etc. It mainly classified into ten categories, namely buildings, different manmade structure, track, road, crops, trees, standing water, waterways, vehicle large, and vehicle small. This dataset also contains the problem of class imbalance.

One of the original images and its mask (labels for each class) are shown in Fig. 4.

Table 1 Description of DSTL Dataset

Wavelength	Dynamic range (bits)	Sensor resolution (meter)	Size
RGB+P (450–690 nm)	11	0.31	3348×3392
M band (400–1040 nm)	11	1.24	837×848
A band (1195–2365 nm)	14	7.5	134×136

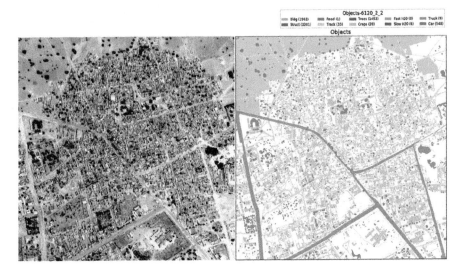

Fig. 4 Original Image and labels for all classes [36]

5.3 Data Transformation

We have used dataset from Kaggle which we will explain later. In this dataset, they created a set of geo-coordinates that are in the range of $x = [0,1]$ and $y = [-1,0]$. Then the coordinates are transformed using the given values Xmax and Ymin for each image. We have width (W) and height (H) from the image. Then we can scale the data as follows:

$$W' = W \cdot \frac{W}{W+1} \qquad (4(4))$$

$$x' = \frac{x}{x_{\max}} \cdot W' \qquad (5(5))$$

$$H' = H \cdot \frac{H}{H+1} \qquad (6(6))$$

$$y' = \frac{y}{y_{\min}} \cdot H' \qquad (7(7))$$

Then scaling back to the original coordinates:

$$x = x' \cdot \frac{x_{\max}}{W'} \qquad (8(8))$$

$$y = y' \cdot \frac{y_{\min}}{H'} \qquad (9(9))$$

5.4 Data Augmentation

U-NET has performed strikingly well on numerous computer vision assignments. Notwithstanding, these models are intensely dependent on huge information to avoid from overfitting. Overfitting is when an application learns a capacity with high fluctuation, for example, to entirely demonstrate the preparation information. Tragically, numerous application spaces do not approach big data, for example, clinical picture investigation. Data augmentation incorporates a set-up of methods that upgrade the size and nature of preparing datasets with the end goal that better deep-learning models can be constructed with their help. These image augmentations also utilized in this work incorporate geometric changes, shading space augmentation, random erasing, kernel filters, adversarial training, neural style transfer, feature space augmentation, meta-learning, and generative adversarial networks. Moreover, data augmentation has many popularly used methods like principal component analysis (PCA) jittering, color jittering, noise, horizontal/ vertical flip, cutting, rotation or reflection, etc. In our dataset, we took images of same resolution so we have used three augmentation methods random crop, horizontal or vertical flip and rotation. We have extracted the image patches of size 160×160 from the original image and then randomly flipped horizontally or vertically.

5.5 Performance Metrics

For the analysis of our model architecture, we used Jaccard Index which is also known as Intersection over Union. It can be interpreted as similarity measure between finite samples of sets. Jaccard score can be calculated as:

5.5.1 Jaccard Index

For the analysis of our model architecture, we used Jaccard Index which is also known as Intersection over Union. It can be interpreted as similarity measure between finite samples of sets. Jaccard score can be calculated as:

$$\text{score} = \frac{1}{n}\sum_{i=1}^{N}\text{jaccard}_i = \frac{1}{N}\sum_{i=1}^{N}\frac{TP_i}{TP_i + FP_i + FN_i} \quad (10(10))$$

where TP is True Positive, FP is False Positive, N is number of classes, and FN is False Negative all are shown in Confusion matric in Fig. 5.

5.5.2 Loss Function

In our case, we have used binary cross entropy as a loss function. For classification tasks, categorical cross entropy is the most common method. But we are using binary cross entropy because in our case classes are not mutually exclusive. An optimization problem aims to minimize a loss function. The formula for binary cross entropy is shown below:

$$H = -\frac{1}{n}\sum_{t-1}^{n}[y\log(\hat{y}) + (1-y)\log(1-\hat{y})] \quad (11(11))$$

5.6 Segmentation Results

Fig. 6 shows the performance matric after training our modified U-net model for 80 epochs iterations on, with batch size equals to 64. We have passed binary cross

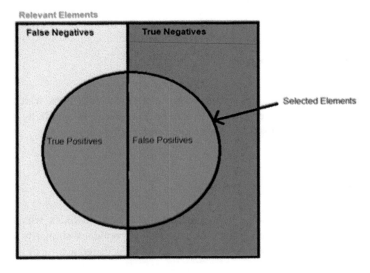

Fig. 5 Illustration of FP, TP, TN, FP

entropy as our loss function and we used Adam optimizer with learning rate 0.0001. We have passed three functions in evaluation matric while model compiling. These are accuracy, Jaccard index as jaccard_coef, rounded predictive value of Jaccard index as jaccard_coef_int. The accuracy, loss, Jaccard index, and rounded predictive value of Jaccard index for both training and validation datasets are depicted in Figs. 6, 7, 8, and 9, respectively.

We executed our model in Kaggle Platform. We have extracted the image patches of size 160×160 from the original image and then randomly flipped horizontally or vertically. The training of model for 80 epochs is shown below in the figure.

We passed three functions in the evaluation matric these are jaccard_coef (Jaccard index), jaccard_coef_int (Rounded value of Jaccard Index), and accuracy. After 80 epochs the value of accuracy is around 97.89% and val_accuracy is 96.01% which is greater than other state of art methods [36].

Our model has been implemented for the segmentation of the satellite images. We have taken six satellite images for identifying tree and crops. Below figure contains the three columns as a satellite image, an image with predicted tree pixel and last one is predicted tree polygon. The same image is implemented on both tree and crop for classification. The pixel is mainly classified on the basis of two classes, i.e., tree class and crop class. When an image is taken as an input, the following model gives a masked set image. Figure 10 shows the tree and crop segmented results.

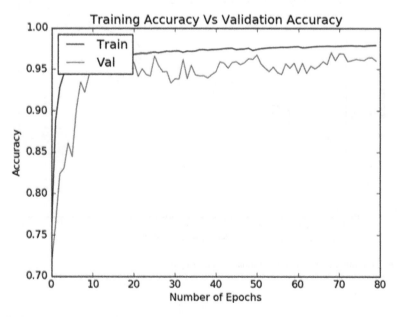

Fig. 6 Training Accuracy Vs Validation Accuracy

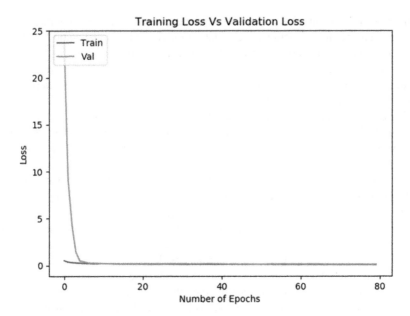

Fig. 7 Training Loss Vs Validation Loss

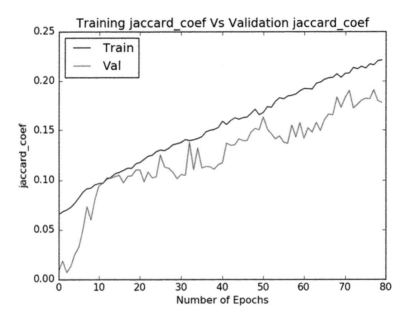

Fig. 8 Training jaccard_coef Vs Validation jaccard_coe

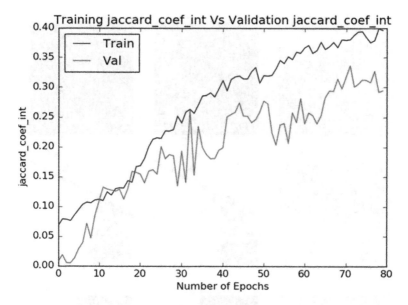

Fig. 9 Training jaccard_coef Vs Validation jaccard_coe

5.7 Analysis

To analyses the performance of proposed model, it has been compared with other state-of-the-art methods, namely LinkNet and ResNet in terms of average Jaccard index. LinkNet [42] is very known network for its focus on predict fast based on the encoder-decoder structure. Well-known problem of loss of spatial information at encoder and not received at decoder is overcome by propagating the spatial information directly from encoder to decoder at corresponding level using LinkNet. Therefore, a significant decrease in processing time because relearning the lost structures are fast and less complex in it. Moreover, LinkNet is unable to deal with the longer-range context information which is a vital dependency for the high-resolution aerial imagery semantic segmentation. While in ResNet [21], nearly 11 million trainable parameters are present. CONV layer consisting of a 3×3 filter is a part of ResNet. All over the network, only two pooling layers, one at the beginning and another at the end of the network are used. Every two CONV layer has an identity connection in between. It mainly used in solving the vanishing gradient problem. The residual block which is the main component is periodic throughout the network.

With the help of data augmentation, the performance of Skip-U-NET model is significantly increased and the average Jaccard index is 0.823 followed by 0.698 and 0.623 for LinkNet and ResNet, respectively. There are two main reasons for the effectiveness of our model. Firstly, the use of skip connections and second is inclusion of data augmentation technique.

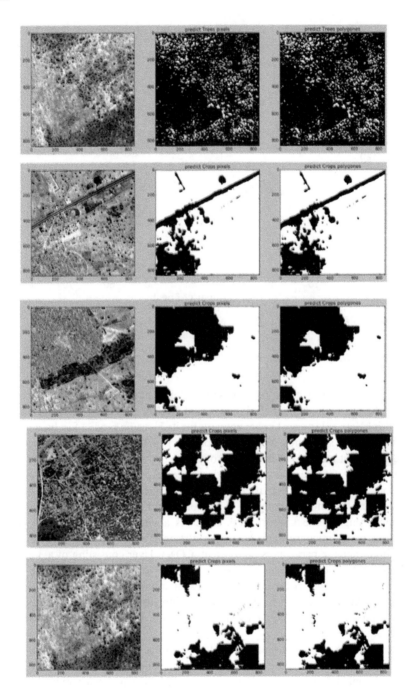

Fig. 10 Segmented results

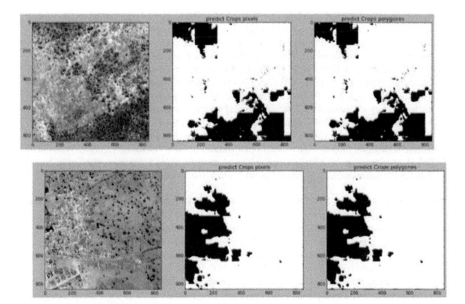

Fig. 10 (continued)

6 Conclusion

U-NET based semantic segmentation is increasingly used in the field of remote sensing, due to its ability to deal with large datasets and its high performance. It is widely used to detect landcover, tree, building, and many other objects. In this chapter, we only considered to detect tree and crop from the satellite images. We also pre-process the data before passing into the model. We have improved the performance of U-NET by hybridizing it with ResNet. After hybridization, the proposed model is given the accuracy of 97.89%. The Skip-U-NET model can further be used in different application areas of image segmentation. In future, we will develop a model which can be trained in weak supervision for the larger applicability of the model.

References

1. Forestier, G., Puissant, A., Wemmert, C., & Gançarski, P. (2012). Knowledge-based region labeling for remote sensing image interpretation. *Computers, Environment and Urban Systems, 36*(5), 470–480.
2. Pal, R., Saraswat, M., & Mittal, H. (2021). Improved bag-of-features using grey relational analysis for classification of histology images. *Complex & Intelligent Systems*, 1–15.
3. Mittal, H., Pandey, A. C., Saraswat, M., Kumar, S., Pal, R., & Modwel, G. (2021). A comprehensive survey of image segmentation: Clustering methods, performance parameters, and benchmark datasets. *Multimedia Tools and Applications, 9*, 1–26.

4. Mittal, H., Pandey, A. C., Pal, R., & Tripathi, A. (2021). A new clustering method for the diagnosis of CoVID19 using medical images. *Applied Intelligence, 51*(5), 1–24.
5. Mittal, H., Tripathi, A., Pandey, A. C., & Pal, R. (2020). Gravitational search algorithm: A comprehensive analysis of recent variants. *Multimedia Tools and Applications, 80*, 1–28.
6. Pal, R., Yadav, S., & Karnwal, R. (2020). EEWC: Energy-efficient weighted clustering method based on genetic algorithm for HWSNs. *Complex & Intelligent Systems, 6*(2), 1–10.
7. Pal, R., & Saraswat, M. (2020). A new weighted two-dimensional vector quantisation encoding method in bag-of-features for histopathological image classification. *International Journal of Intelligent Information and Database Systems, 13*(2–4), 150–171.
8. Pal, R., & Saraswat, M. (2019). Histopathological image classification using enhanced bag-of-feature with spiral biogeography-based optimization. *Applied Intelligence, 49*(9), 3406–3424.
9. Pal, R., & Saraswat, M. (2019). Grey relational analysis based keypoints selection in bag-of-features for histopathological image classification. *Recent Patents on Computer Science, 12*(4), 260–268.
10. Girshick, R., Donahue, J., Darrell, T., & Malik, J. (2014). Rich feature hierarchies for accurate object detection and semantic segmentation. *Proceedings of the IEEE conference on computer vision and pattern recognition*, Columbus, OH, 2014 (pp. 580–587).
11. Garcia-Garcia, A., Orts-Escolano, S., Oprea, S., Villena-Martinez, V., & Garcia-Rodriguez, J. (2017). A review on deep learning techniques applied to semantic segmentation. *Transactions on Pattern Analysis and Machine Intelligence, 0*, 0. https://arxiv.org/abs/1704.06857
12. Pal, R., Mittal, H., & Saraswat, M. (2019, November). Optimal fuzzy clustering by improved biogeography-based optimization for leukocytes segmentation. In *2019 Fifth International Conference on Image Information Processing (ICIIP)* (pp. 74–79). IEEE.
13. Application of Semantic segmentation - https://www.imageannotation.ai/blog/semantic-segmentation-applications
14. Zhong, Y., Fei, F., Liu, Y., et al. (2017). SatCNN: Satellite image dataset classification using agile convolutional neural networks. *Remote Sensing Letters, 8*(2), 136–145.
15. Senthilnath, J., Kandukuri, M., Dokania, A., & Ramesh, K. N. (2017). Application of UAV imaging platform for vegetation analysis based on spectral-spatial methods. *Computers Electronics in Agriculture, 140*, 8–24.
16. Hu, Q., et al. (2013). Exploring the use of Google earth imagery and object-based methods in land use/cover mapping. *Remote Sensing, 5*, 6026–6042.
17. Simonyan, K., & Zisserman, A. (2014). Very deep convolutional networks for large-scale image recognition. arXiv Preprint ArXiv, *1409*, 1556.
18. Marmanis, D., Datcu, M., Esch, T., & Stilla, U. (2016a). Deep learning earth observation classification using ImageNet pretrained networks. *IEEE Geoscience and Remote Sensing Letters, 13*(1), 105–109.
19. Penatti, O. A. B., Nogueira, K., & Santos, J. A. D. (2015). *Do deep features generalize from everyday objects to remote sensing and aerial scenes domains? Computer vision and pattern recognition workshops*, IEEE, Boston, MA, 2015 (pp. 44–51).
20. Kondrich, A., Kasevich, I.. (2018). TreeNet: Deep U-Net for Image Segmentation. Stanford University, CA, CS230: Deep Learning, Spring.
21. He, K., Zhang, X., Ren, S., & Sun, J. (2016). Deep residual learning for image recognition. In *Proceedings of the IEEE conference on computer vision and pattern recognition* (pp. 770–778).
22. Mittal, H., Saraswat, M., & Pal, R. (2020, January). Histopathological image classification by optimized neural network using IGSA. In *International conference on distributed computing and internet technology* (pp. 429–436). Springer.
23. Decatur, S. E. (1989). Application of neural networks to terrain classification. *International joint conference on neural networks, IEEE*, (Vol. 1, 1989, pp. 283–288).
24. Zhang, L., Zhang, L., & Du, B. (2016). Deep learning for remote sensing data: A technical tutorial on the state of the art. *IEEE Geoscience and Remote Sensing Magazine, 4*(2), 22–40.
25. Bischof, H., Schneider, W., & Pinz, A. J. (1992). Multispectral classification of Landsat-images using neural networks. *IEEE Transactions on Geoscience and Remote Sensing, 30*(3), 482–490.

26. Mountrakis, G., Im, J., & Ogole, C. (2011). Support vector machines in remote sensing: A review. *ISPRS Journal of Photogrammetry and Remote Sensing, 66*(3), 247–259.
27. Pal, M. (2005). Random forest classifier for remote sensing classification. *International Journal of Remote Sensing, 26*(1), 217–222.
28. Atkinson, P. M., & Lewis, P. (2000). Geostatistical classification for remote sensing: An introduction. *Computers & Geosciences, 26*(4), 361–371.
29. Han, J., Zhang, D., Cheng, G., Guo, L., & Ren, J. (2015). Object detection in optical remote sensing images based on weakly supervised learning and high-level feature learning. *IEEE Transactions on Geoscience and Remote Sensing, 53*(6), 3325–3337.
30. Mnih, V. (2013). *Machine learning for aerial image labeling* (p. 2013). University of Toronto.
31. Shelhamer, E., Long, J., & Darrell, T. (2017). Fully convolutional networks for semantic segmentation. *IEEE Transactions on Pattern Analysis and Machine Intelligence, 39*(4), 640–651.
32. Garcia-Garcia, A., Orts-Escolano, S., Oprea, S., Villena-Martinez, V., & Garcia-Rodriguez, J. (2017). A review on deep learning techniques applied to semantic segmentation. *Transactions on Pattern Analysis and Machine Intelligence, 0*, 0. https://arxiv.org/abs/1704.06857
33. Ivanovsky, L., Khryashchev, V., Pavlov, V., & Ostrovskaya, A. (2019, April). Building detection on aerial images using U-NET neural networks. In *2019 24th Conference of Open Innovations Association (FRUCT)* (pp. 116–122). IEEE.
34. Ghosh, A., Ehrlich, M., Shah, S., Davis, L., & Chellappa, R.. (2018). Stacked U-Nets for Ground Material Segmentation in Remote Sensing Imagery. In *2018 IEEE/CVF Conference on Computer Vision and Pattern Recognition Workshops (CVPRW)*, Salt Lake City, UT, 2018 (pp. 252–2524). https://doi.org/10.1109/CVPRW.2018.00047
35. Iglovikov, V., Mushinskiy, S., & Osin, V. (2017). Satellite Imagery Feature Detection using Deep Convolutional Neural Network: A Kaggle Competition. arXiv:1706.06169.
36. Zhihuan, W., Gao, Y., Li, L., Xue, J., & Li, Y. (2019). Semantic segmentation of high-resolution remote sensing images using fully convolutional network with adaptive threshold. *Connection Science, 31*(2), 169–184. https://doi.org/10.1080/09540091.2018.1510902
37. Jasmine, H., Rai, M., & Wong, V. (2018). *Building detection from satellite imagery*. Stanford University.
38. Culberg, Kevin, & Fuhs, K. (2017). *"Feature Extraction in Satellite Imagery Using Support Vector Machines."*
39. Khryashchev, V., Ivanovsky, L., Pavlov, V., Ostrovskaya, A., & Rubtsov, A. (2018). Comparison of Different Convolutional Neural Network Architectures for Satellite Image Segmentation. In *2018 23rd Conference of Open Innovations Association (FRUCT)*, Bologna (pp. 172–179). https://doi.org/10.23919/FRUCT.2018.8588071
40. Avenash, R., & Viswanath, P. (2019). *Semantic segmentation of satellite images using a modified CNN with hard-swish activation function,* In Proceedings of the 14th International Joint Conference on Computer Vision, Imaging and Computer Graphics Theory and Applications, India, (pp. 413–420). https://doi.org/10.5220/0007469604130420
41. Osin, V., Cichocki, A., & Burnaev, E. (2018). Fast multispectral deep fusion networks. *Bulletin of the Polish Academy of Sciences: Technical Sciences*, 875–889. https://doi.org/10.24425/bpas.2018.125935
42. Chaurasia, A., & Culurciello, E. (2017, December). LinkNet: Exploiting encoder representations for efficient semantic segmentation. In *2017 IEEE Visual Communications and Image Processing* (VCIP) (pp. 1–4). IEEE.

Artificial Cyber Espionage Based Protection of Technological Enabled Automated Cities Infrastructure by Dark Web Cyber Offender

Romil Rawat, Vinod Mahor, Sachin Chirgaiya, and Bhagwati Garg

1 Introduction

Simulated intelligence and machine learning (ML) have become basic advances in data security, as they can rapidly dissect a huge number of occasions and recognize a wide range of sorts of dangers—from malware misusing zero-day weaknesses to distinguishing hazardous conduct that may prompt a phishing infringement or download of pernicious code. The moderated technologies based on AI [1, 2] designed to protect from suspicious activities and infringements, for generating parameters and signature based authentication procedures. ML works best when focused on a particular undertaking instead of a wide-running mission. Master frameworks are programs intended to tackle infringement issues inside particular spaces. By impersonating the considering human specialists, they tackle issues and settle on choices utilizing fluffy standards-based thinking through cautiously curated groups of statistics. Neural organizations utilize a naturally aroused programming worldview which empowers to gain from observational statistics. In a neural organization, every hub relegates a load to its statistics addressing how right or off base it is comparative with the activity being performed. The last yield is then controlled by

---

R. Rawat (✉) · S. Chirgaiya
Department of Computer Science Engineering, Shri Vaishnav Vidyapeeth Vishwavidyalaya, Indore, India

V. Mahor
Department of Computer Science and Engineering, IPS College of Technology and Management, Gwalior, India

B. Garg
Union Bank of India, Gwalior, India

© The Author(s), under exclusive license to Springer Nature Switzerland AG 2021
F. Al-Turjman et al. (eds.), *Intelligence of Things: AI-IoT Based Critical-Applications and Innovations*,
https://doi.org/10.1007/978-3-030-82800-4_7

the amount of such loads. Deep learning [3, 4] (DL) is important for a more extensive group of ML strategies dependent on learning statistics portrayals, instead of errand explicit calculations. Today, picture acknowledgment through DL is regularly in a way that is better than people, with an assortment of utilizations, for example, self-governing vehicles, check examinations, and clinical conclusions. The generations of audio-visual data framed by techniques Aretha open target of vulnerability. Infiltrator may represent someone who says or does almost anything using ever-decreasing quantities of initial content. But the possibility of such illegal uses of deep forged [5] is rising as technology progresses. In December, Facebook revealed that a pro-Trump US media [6] outlet used similar technologies to create images for hundreds of bogus profiles that were then used to push political posts. It could just as easily be used to create special avatars to mask the identity of infiltrator when constructing scams. Groups like Deep Trace are working on technologies to reveal these synthetic images, invader have the ability to get the data pretty much all the residents inside only minutes bringing about an innate hazard to their prosperity. A mechanical city additionally distributes open statistics, for example, data to the resident about their city. It has affected us in a manner that is valuable to utilize and is simply going to ameliorate the personal satisfaction for all general population in these mechanical empowered states. In this part we will bring up the surveillance issues of mechanical city digital framework, with the end goal that the residents can totally understand the advantages that the innovation can offer. The automated chip enabled [7] components using advance learning algorithms [8, 9] are working to control human [10, 11] thinkings, the below Fig. 1 shows about the Brain–Algorithms Intelligence Interfacing.

1.1 Using AI to Prevent Cybercrime Targeting Technological Enabled Colonies on the DW

The surface of the cyberspace used by ordinary populace is a deep DW where underground markets flourish in the sale of illicit guns, drugs, credit card numbers, counterfeit credits, and computer hacking tools. With increased dependency on automated technologies such as Artificial Intelligence (AI), legislation enforcement (LE) [12] authorities have worked continuously to avoid the disruptive acts of offenders who believe they are safe behind the screen of anonymity. Extracting cyber espionage [13] from deep dark net browsers can be cumbersome, and for offender or terrorists attempting to mask their previous offender background, discovering the suspect's digital fingerprints that will link the dots and reveal inherent clues is time-consuming. Technology built with artificial cyber espionage-driven search engines [14] that are capable of shifting through unlimited volumes of vital data and methodologies that satisfy the criteria of LE agencies across the globe helps authorities to crack down on hidden web offender enterprises in a minimum period

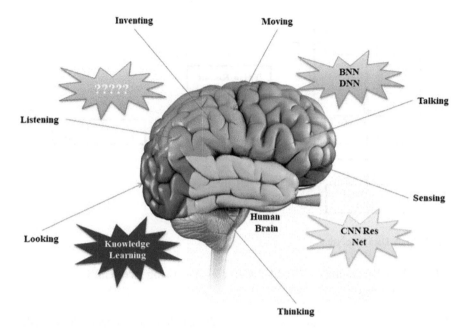

Fig. 1 Brain–Algorithms Intelligence Interfacing

of time with minimal resources. Much as authorities are using those technologies against offenders online, technological enabled colonies are using AI and other modern technologies to eradicate violence and improve security in urban frameworks. Cities are increasingly using AI and Visual Recognition Technologies to search and recognize unusual activities that may signify inherent illegal or terrorist [15] activity, as well as to use face recognition technology to identify suspects who could be sought by the authorities. Automated AI [16, 17] web cyber espionage processing engines can unveil the identity of offenders using face recognition technologies, identify human faces and features in photos across various tiers of the web, and produce real-time notifications for faces that appear to be similar. Yet even with the many advantages, technological enabled colonies [18] may provide infringement of privacy and inadequate computer protection, which could unintentionally risk data from populace falling into the hands of cyber offender. Huge volumes of data contained in technological enabled colonies are vaults waiting to be cracked by infiltrator who are actively developing and selling new malware on hidden web browsers. Though illegal activity is rampant in the deep DW [19], it is also a rich source of cyber espionage. Using artificial cyber espionage technologies, officials are alerted to real-time actionable warnings that can deter devastating illicit filtration as well as full peril cyber espionage visual from the DW that will allow them to further plan or mitigate inherent illicit filtration. The brain data and facts are gathered using intelligence techniques and processed for decisions generation. Fig. 2 outlines about the Brain Data Processing mechanism.

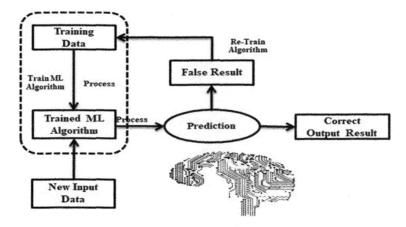

Fig. 2 Brain Data Processing

1.1.1 How Technological Enabled Colonies Are Using Artificial Cyber Espionage

There are different stages in which AI (ML) [20] can have an influence. Man-made intelligence permits geospatial properties to be gotten from guides and area following propriety in a programmed way. Anyway, it is significant different fields that fall after this first phase of extraction, for example, the investigation of spatial-transient metropolitan [21] patterns (metropolitan development, framework organization, versatility and neighborhood relations), with geospatial designs upgraded with trait metadata [22], for example, building use subtleties, transport plans, AI might be utilized to assess patterns and heterogeneities in the district. There are different stages in which ML can have an influence. First in the feeling of Earth Observation, AI permits geospatial qualities to be gotten from maps and Geographic data Web-Modeling [23] propriety in a programmed way. Anyway, it is important different fields that come after this first phase of extraction with geospatial designs improved with property metadata. Fig. 3 outlines about the modeling of Smart Cities Models.

1.1.2 What AI and ML can do for Automated City?

Artificial cyber espionage and deep learning algorithms [24] have rapidly become an important part of a variety of sectors. They are also finding their way to automated city projects with the goal of automating and advancing urban processes and operations in general. Usually, a city known as automated city means that it is using some sort of IoT and ML [25] to collect data from different points. Technological enabled colonies have different appropriateness of AI-driven and IoT-enabled technologies [26], from ensuring a cleaner atmosphere to supporting public transport and safety. By using AI and ML algorithms, along with IoT [13], a

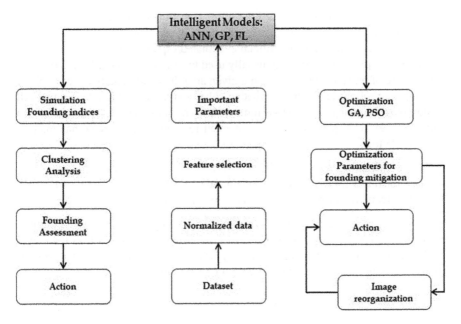

Fig. 3 Smart Cities Models

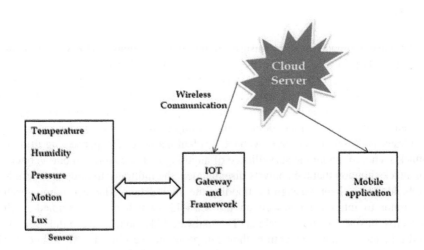

Fig. 4 Wireless communication gateway framework

community will plan automated traffic strategies to ensure that populace get as safe and effective as possible from one point to the next. ML gathers data from a variety of points and forwards it to the central server for further deployment, and once data is obtained, it must be used to make the city automated. Fig. 4 outlines about the wireless communication gateway framework.

It is not restricted to city-related activities, yet works well in fisheries, ranger service, philanthropic help—anything is possible, where we focus on how technological enabled colonies utilize artificial cybernated espionage [27, 28].

Automated plans are frequently usually used to follow solar boards and to gauge the measure of inexhaustible power in a given area. It is additionally conceivable to contemplate rooftops and recreate whether solar boards can be mounted on them. It is additionally simple to uncover swimming shower, for instance, for geo-advertising purposes [29], or to perform charge statement reviews. The interface has made it conceivable to:

- Reveal swimming shower or different sorts of swimming shower;
- Place relating with the package speaking to,
- Build advertising or review unit perception entrance.

Rising networks are confronted with the undertaking of having sufficient burial ground space for their populaces. ML is a ground-breaking answer [30] for graveyard administration, as it permits assessing where and when more burial grounds should be set up. Graves can be found with only a couple taps.

2 WEBNET Can Be Hacked Through Automated Light Bulb

Cybernated wrongdoer could misuse an Internet of Things (IoT) webnet—automated lights [31] and their control connectto dispatch illegal filtration on regular system webnet [32] connected using automated utilities. The later equipment ages of Hue enlightenment bulb [33] do not have the misused weakness, IoT contraption can represent a surveillance hazard, however this exploration shows how even the most ordinary, apparently idiotic device, for example, light bulb can be abused by infiltrator and used to take over webnet, or plant malware. The perplexing fifth-age attack scene, to ignore the surveillance of anything that is coupled to our webnet in an attack situation that the analysts disentangled, the infiltrator to find the loopholes embedded into client's design for flooding the illicit details. The facts empower the infiltrator to introduce malware on the scaffold—which is thus coupled to the objective business or home webnet. The malware [34] likens back to the infiltrator and utilizing a known adventure, they can penetrate the objective IP webnet [35] from the extension to spread ransomware or spyware [36].

2.1 Automated City Surveillance Issues

Portion of significant surveillance issues related with the automated metropolis surveillance, quick development of the advances empowering the idea of automated cities could surveillance concerning those advances can be kept up. New contraption, for example, tablets, workstations, automated telephones, and more have made it simpler for the inalienable intruder to discover openings in the cybernated foundation. Additionally, with the presentation of wireless enabled transmission there is a consistent cybernated space access whenever in specific urban areas [37], risk level for an ambush to happen has just expanded another connected issue is the preparation of workers who really know how to impregnable automated metropolis webnet. With such quick development and extension of technological enabled colonies [38, 39] there are not many protections skilled that have the capability to real keep up and uphold automated metropolis surveillance Web-Modeling [40]. The automated metropolis surveillance field is at present so understaffed that it is required to get one of the best five generally pursued and needed positions sooner rather than later regarding the innovation market. Prepared people in this field are sought after. Without affirmed people, it is hard to speaking to the surveillance issues. Deficiency of the surveillance capable makes it simple for the trespasser to recognize and target more openings in the webnet so that danger transforms into an attack. Another issue is the fix arrangement and surveillance refreshes. For instance, for each new update that happens, there will be some new sort of surveillance [41] opening that will open up in the cybernated foundation. Additionally, with such fast development of automated metropolis activities, the updates are hurried out rapidly. Therefore, there will be openings for an ambush, for example, a SQL infusion. Here are a few instances of ramifications of straightforward issues in the code or if an ambush happens on the automated metropolis surveillance webnet. Another deterrent that numerous general populations do not consider and it has a colossal compare particle to technological enabled colonies is the financial plan for technological enabled colonies project [42]. These technological enabled colonies are not being paid for by one individual or a gathering of people. The cash that is paying for these technological enabled colonies is coming from the metropolis itself meaning the duty currency that the urban communities gather. With the steady changing spending circumstances and how it is influence things like instruction, social projects, etc. The financial plan for a metropolis has an immediate connection to an automated project since it directs how much cash a metropolis should spend on their automated surveillance exercises [43]. Ordinarily whatever gives a metropolis is confronting will be featured in the media and other such issues. The financial plan of that metropolis will be moved towards a bigger level of that spending plan going towards that and different things that the metropolis is paying for will get less. As has been talked about so far, an automated power center points nearly all that is running the everyday needs of the metropolis. For example, automated metropolis runs things like traffic cameras, water and sewage lines, and electrical plants, etc. In the event that the proper spending plan is not given to the people that are running these technological enabled colonies they will not have the option to satisfactorily secure these urban areas. A spending plan can be a quiet enemy of automated metropolis surveillance activity. Without having the legitimate financial plan, it can prompt circumstances [44], for example, absence of prepared and guaranteed surveillance able just as not having the appropriate assets in line to sufficiently ensure an automated metropolis. These can go from both equipment and programming needs. For an automated metropolis surveillance activity, it is critical

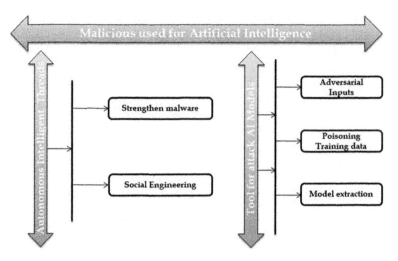

Fig. 5 Cyber Threats Intelligence framework

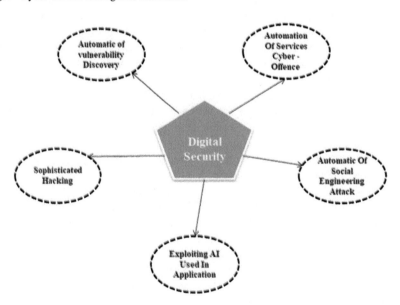

Fig. 6 Digital security framework

these messages are transferred to the arrangement creators, managers and ensure the issues of a continually changing spending plan are known to these pioneers. Without a set spending that is talked about and arranged out it can leave automated metropolis open to more hazard and unlawful filtration then many would think to be conceivable. Figure 5 below shows about the Cyber Threats Intelligence framework and Fig. 6 below outlines about the digital security framework architecture.

2.2 Seclusion and Surveillance

Smart Colonies gather enormous volumes of statistics utilizing cameras and sensing chips for various suitability, for example, traffic thinking about and security. Because of forceful statistics assortment, a few social liberties activists and scientists have worries about the ambush of seclusion. Steady camera [33] surveillance will prompt the foundation of a surveillance state. Additionally, ceaseless statistics assortment helps government specialists see each part of their resident's lives. Information surveillance is another significant issue that is related with gathering information for technological enabled colonies. Huge information vaults are inclined to different cybernated unlawful filtrations. Cybernated wrongdoer is continually planning new malware and cybernated unlawful filtration to acquire illicit admittance to resident information. Intruders are continually attempting to make stealthier and more perilous cybernated unlawful filtration [35]. For example, in the metropolis of Del Rio, Texas, a ransomware ambush locked down numerous workers in Metropolis Hall. All administration tasks were taken care of utilizing pen and paper alongside no admittance to verifiable records. Such cybernated hazard can be tragic for any automated metropolis and represents a significant risk to resident information. Figure 7 outlines for the Cyber Security Threat Vectors.

2.3 Information Bias

Subsequent to gathering enormous volumes of information, governments produce instructive investigation to acquire bits of knowledge into various parts of

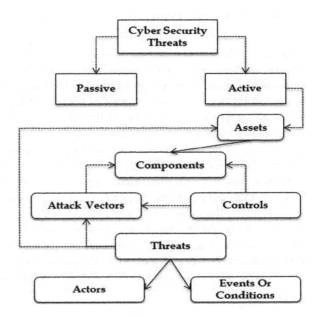

Fig. 7 Cyber Security Threat Vectors

technological enabled colonies utilizing large information examination. The information used for investigation [45] is authentic information and ongoing information. Such information may incorporate human inclinations prompting incorrect outcomes. Preparing engines using this information may unexpectedly victimize explicit religions, identities, or sexes. For example, an understudy assessment Web-Modeling for school contains information that is one-sided against a particular race. At that point, the Web-Modeling will convolute the endorsement cycle for understudies. The inadvertent predisposition can prompt desperate ramifications for a few people. The instance of information inclination brings up an enormous issue of responsibility. The wake of recognizing inadvertent information predisposition, finding the defective information [20] among the enormous volume of information can be a "needle in the sheaf" circumstance. For this situation, erasing all the gathered information and beginning once again with exact, fair information is by all accounts the lone alternative. Yet, invigorating the whole assemblage of information will be costly and incredibly tedious. Henceforth, technological enabled colonies ought to be cautious while choosing information hotspots for automated metropolis suitability. A great many connected "things" are introduced in technological enabled colonies around the globe. The rise of the IoT uncovers an expansive assortment of shortcomings that cybernated trespasser [40] and other noxious entertainers may use. While technological enabled colonies are planned to improve seriousness and execution, they can inalienably introduce huge dangers to residents and specialists when information surveillance is disregarded. There are an undisclosed number of potential weaknesses and philosophies, the absolute most incessant illegal filtration includes:

3 Solving Urban Issues with AI and ML

IT typically takes data created by a variety of appropriateness, cyberspace-enabled vehicles and leverages it to reveal trends and learn how to improve a given range of amenity. Its resources are able to personalize the automated metropolis experience by aggregating details on the most traveled roads in the metropolis and then adding it to the transportation infrastructure. In the other hand, ML and AI will aid in the processing of waste and its proper governance and disposal, which is a critical civic operation in a region. Automated recycling and waste governance processing engines also offer a safe waste governance system. AI has the opportunity to understand how cities are being used and how they work. It lets officials to consider different shifts and programs. AI-powered computer vision processing engines, for example, may allow computers to spot millions of elements of urban life in a chorus, like individuals, public employees, vehicles, incidents, explosions, hazards [31], garbage, and much more. The framework allows not only autonomous control, but also decisions to be made on the basis of the output of each of these components, the shift in actions over the course of each day or period, and the reaction of each

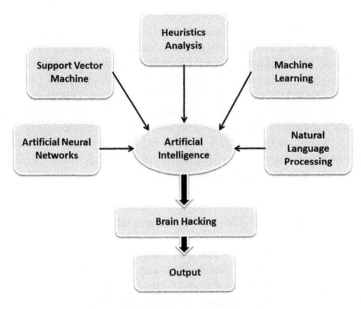

Fig. 8 Brain Hacking Technological Designs

element to metropolis processing engines. Fig. 8 shows about the Brain Hacking Technological [15] Designs.

As AI and ML improve the way communities work, provide, and sustain public amenity, innovations come with certain pitfalls. There is also a need to explore retrofitting options that will keep automated metropolis projects running. As a result, new automated metropolis projects using AI and ML tend to advance metropolis amenity and life, including transport, lighting [23], defense, webnet, health amenity, among others.

3.1 Technological Enabled Colonies Peril

Technological enabled colonies are the embedded chip programming solutions for easiness with security authorities and modeling. Few frequent illicit filtrations include:

3.1.1 Perpetrator Eavesdrop Threat

A trespasser parts, postponements or satires correspondences between two gadgets. A man-in-the-medium ambush on cybernetic valve in a wastewater Web-Modeling might be utilized to cause a biohazard spill. An eavesdropping penetrates, otherwise called sneaking around or sniffing, is an organization security infringement where an individual attempts to take the dossier that cell phones and other advanced gadgets send or get this hack profits by unstable organization transmissions to get to the statistics being sent. Listening in is trying to distinguish since it does not cause strange statistics transmissions. These infringements target debilitated transmissions between the customer and worker that empower the aggressor to get web-mesh transmissions. An aggressor can introduce web-mesh screens, for example, sniffers to play out a listening in infringement and capture statistics as it is being sent. Any gadget inside the sending and accepting organization is a weakness point, including the terminal and beginning gadgets themselves. One approach to ensure against these infringements understands what gadgets are associated with a specific organization and what programming is run on these gadgets.

3.1.2 Dossier and Identity Thievery by Gadget Hijacking

Statistics made by unregulated cybernetic metropolis framework, for example, parking structures, E-charging stations [31], and surveillance takes care of give cybernated trespasser a wide assortment of focused individual subtleties that can naturally be utilized for fake buys and to distinguish robbery. At the point when a criminal deceitfully recognizes themselves to police as one more individual at the purpose of capture, it is now and again alluded to as "Criminal Identity Theft." At times, lawbreakers have recently acquired state-gave character records utilizing certifications taken from others, or have basically introduced a phony ID. Given the deception works, charges might be set under the casualty's name, letting the criminal free. Casualties may just learn of such episodes by some coincidence, for instance, by accepting a court summons, finding their drivers licenses are suspended when halted for minor petty criminal offenses, or through historical verifications performed for business purposes. It tends to be hard for the survivor of a criminal wholesale fraud to clear their record. The means needed to clear the casualty's mistaken criminal record depend on which purview the wrongdoing happened and whether the genuine character of the criminal can be resolved. The casualty may have to find the first capturing officials and demonstrate their own character by some solid methods, for example, fingerprinting or DNA testing, and may have to go to a court hearing to be found not guilty. Getting an expungement of court records may likewise be required. Specialists may forever keep up the casualty's name as a nom de plume for the criminal's actual personality in their criminal records statistic's bases. One issue that survivors of criminal fraud may experience is that divergent statistics aggregators may in any case have the mistaken criminal records in their dossier sets even after court and police records are amended. Consequently, it is conceivable that a future individual verification will restore the wrong criminal records. This is only one illustration of the sorts of effect that may keep on influencing the casualties of wholesale fraud for certain months or even a very long time after the wrongdoing, beside the mental injury that being "cloned" commonly incites. The interloper seizes and adequately claims the system. This illegal filtration can be hard to uncover on the grounds that by and large, the trespasser does not change the center marshaling of the Web-Modeling. In the feeling of a cybernetic metropolis, cybernated wrongdoer may utilize hacked cybernetic power utilization analyzer to release ransomware illegal filtration on Energy Governance Processing engines.

3.1.3 Network Information Flooding Threat

It plans to make a PC or webnet asset distant to its expected clients by crippling the pleasantry of a host compared to the cybernated space briefly or forever. This is generally cultivated by overpowering the objective with undesirable requests to maintain a strategic distance from the satisfaction of legitimate solicitations. On account of a circulated (DDoS) [46], the approaching traffic floods come from various destinations, making it difficult to dodge the cybernated hostile by just hindering a solitary source. Inside technological enabled colonies, a plenty of handling engines, for example, stopping controlling chips, could be penetrated and constrained to enter a Botnet Intended to Over-Burden the Foundation by Simultaneously Requesting a Liturgy. Permanent Denial of Assistance (PDoA) likewise referred to freely as plashing is an attack that influences the Web-Modeling so seriously that it needs equipment fix or reinstallation [41]. In an automated metropolis model, a commandeered stopping controlling chips may even be the objective of defacement and would need to be destroyed. Below are the featured security modeling for smart automations.

Securing Technological Enabled Colonies

Connected automated metropolis device ought to be impregnable with powerful IoT assurance arrangements (contraption to cloud). Viable and straightforward, yet protected, arrangements that can be rapidly and broadly executed by Original Equipment Manufacturer [37] and comfort are more effective than super arrangements that do not accomplish critical foothold. These innovations ought to have the accompanying abilities.

Firmware Integrity and Impregnable Boot

Safe boot utilizes cryptographic code marking strategies to guarantee that a Web-Modeling just executes code contraption. The utilization of impregnable boot advances prevents infiltrator from trading firmware with malevolent forms and keeps unlawful filtration [36] from happening. Tragically, not all IoT chipsets are fitted with impregnable boot ability. In such a case it is important to guarantee that the IoT Web-Modeling can just liken with affirmed enhancement to diminish the chance of supplanting firmware with malevolent guidance sets.

Mutual Testament

Any time an automated metropolis Web-Modeling [19] compares to a webnet, it ought to be confirmed before getting or communicating information. This implies that the information come from an authentic design and not from a fashioned source. Steady, shared confirmation where two people (device and liturgy) need to demonstrate their personalities to one another assists with safeguarding against vindictive illegal filtration.

Surveillance Contemplating and Analysis

The information is then examined to distinguish suspected features for surveillance and illegal filtration. When noticed, a full assortment of measures detailed in the feeling of the overall surveillance convention of the Web-Modeling [45] ought to be done, for example, isolate device dependent on dubious conduct.

Surveillance Lifecycle Governance

The Lifecycle Contemplating encourages illicit suppliers to follow the parts of IoT device during utilization. Quick over-the-air Web-Modeling key substitution during cybernated emergency recuperation ensures least liturgy interference. Moreover, the safe decommissioning of the Web-Modeling implies that the rejected machines cannot be duplicated and used to compare to the foundation without approval.

Root of Credible

Arrangements offer an equipment level reason for surveillance enabled usefulness, for example, impregnable booting, and impregnable program execution, alter discovery and security, and impregnable [26] key stockpiling and taking care of.

Protocol Engines

Stable webnet Linked between IIOT device and the cloud without modifying traffic. They permit safe start to finish correspondences while holding basic webnet execution.

Provisioning and Key Governance

Safe production network answers for chip and IoT Web-Modeling providers incorporate impregnable key inclusion and cloud-based device key administration enhancement. A few weaknesses and risk face the cybernated-physical webnet [30] utilized in metropolitan automated. Anyway, these new cybernated-actual framework innovations are usually utilized, there is no acceptable understanding into their shortcomings and dangers. When all is said and done, intentional and unintended dangers to automated metropolis framework related assurance cause contrasting huge ramifications dependent on the metropolis's refinement and automated. Accordingly, we are introducing the key hindrance and dangers to foundation identified with protection. Metropolitan utilities, for example, power age, water conveyance, roadways, houses, and others represent a scope of surveillance hazards in their interesting cybernated-actual parts and structures, for example,

- **Cameras:** Cities are brimming with private and public cameras, which are all ensured by encryption.
- **Seclusion and insurance of username/secret key.** Coming to and seeing private or public cameras are a penetration of the privileges of people and espionage of government interests.
- **Correspondence webnet:** Cybernated-actual articles are wired together in an automated metropolis utilizing different correspondence preparing engines, for example, Wi-Fi, 4G, RFID [27], GSM [15] and others. Of them have extraordinary surveillance [20] gives that should be deciphered through the usage and use of correspondence innovation.
- **Building administration handling engines:** Designers and draftsmen of such suitability regularly center on conveying offices and overlook concerns identified with cybernated surveillance. In this way, producers of such handling engines do not keep up these preparing engines with notice choices to caution clients of surveillance breaks and do not respond to weaknesses that bring about perilous and feebly impregnable structure administration preparing engines.
- **Transport administration preparing engines:** These handling engines are confronted with the main hacks as they cause catastrophes, especially on account of air traffic handling engines or train control handling engines. It make significant traffic postpones that can keep going for quite a long time by hacking and sequencing traffic signal control handling engines, street signs, and speed limit signs. Fig. 9 given below shows about the Smart Cities Security Network [10].

The IOT design comprised of different degrees of IOT [46] uphold advancements. It assists with exhibiting how different innovations liken to one another and to pass on the adaptability, measured quality, and marshaling of IOT executions in various situations. The usefulness of every level is characterized as follows:

4 IOT Architectural View

The IOT architecture consists of various levels of IOT support technologies. It helps to demonstrate how various technologies equate to each other and to convey the scalability, modularity, and marshaling [13] of IOT implementations in different scenarios. The functionality of each tier is defined as follows:

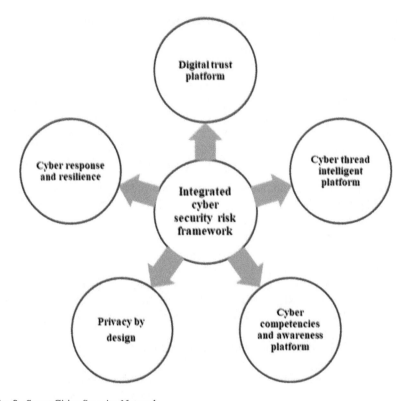

Fig. 9 Smart Cities Security Network

4.1 Automated Gadget/Sensing Chips Tiers

The most minimal levels comprised of automated articles consolidated with the sensing chips. The sensing chips permit the physical and computerized domains to be incorporated, empowering data to be assembled and investigated progressively. Several sensing chips evaluate the live environmental attributes permitting them to log an assortment of estimations. Sensing chips can ascertain the actual property and change it to a sign that can be deciphered by an instrument. Sensing chips are grouped by their particular capacity, for example, structural sensing chips, body sensing chips, home apparatus sensing chips and auto telemetry sensing chips, and so on.

4.2 Gateways and Webnet

These little sensing chips can produce a lot of information, requiring a steady and elite wired or remote webnet framework as methods for transport. Present webnet frequently coupled to vibrant conventions, for assorting technological enabled locations. The fast value-based enhancement, setting mindful propriety, and so on, a few webnet with various advancements and access conventions are needed to map with heterogeneous marshaling. This webnet might be as private, public, or half-breed framework and is designed to help compare standards for inertness, data transfer surveillance.

4.3 Governance Liturgy Tiers

The administration liturgy makes it conceivable to handle data through examination, access controls, measure improvement and Web-Modeling administration. Market and cycle rule engines are one of the fundamental parts of the administration liturgy levels. IOT integrates the connection and association among articles and structures to give data as episodes or spatial facts, for example, heat of things [47, 48], current area, and traffic facts. Any of these occurrences require separating post-handling webnet, for example, successive sensing chips facts, expect reaction to earnest conditions, reacting to understanding well-being crises. Rule engines help the plan of choice rationale and initiate intelligent and automated cycles [31] to permit a more adaptable IOT structure.

4.4 IOT Conceptual View

The key tasks of this system are to evaluate and classify the automated behaviors of these automated gadgets by establishing a complexity between them. The proposed architecture would help to standardize IoT infrastructure so that it can receive e-amenity dependent on qualitative accomplishment, leaving the existing infrastructure unchanged. The successful collaboration of these heterogeneous gadget and protocols will contribute to future ambient computing where the optimum efficiency of cloud computing is assured. This model [26] is capable of logically separating the location of physical gadget, establishing virtual equate ions between various domains, webnet, and coordinating between multiple appropriateness without any central coordination structure. The service is typically delivered by modern data centers operated by giant corporations and consortium. It is defined as resource virtualization that allows the user to deploy and run the program over a virtualization tier and allows the gadget to be distributed, configurable, and scalable.

4.4.1 Equate Tiers

The inherent cyberspace relies to a large degree on the convergence of these familiar objects located anywhere close to us, and they should be easily recognizable and controllable.

This tier also includes the assignment of low-level webnet equipment such as sensing chips, actuators, identifiers, etc. and resource governance tests [30] the availability of physical resources for all gadget and webnet participating in the underlying infrastructure. This gadget contains very small resources and the control of resources guarantees optimum usage with no overhead. It also enables the exchange and dissemination of facts between different webnet and a single webnet separated into multiple domains.

4.4.2 Access Tiers

Context Data can be accessed via IoT Gateway to the Cyberspace as collected by short-range gadget in the form of raw data. Access tiers require topology specification, webnet initialization, domain formation, etc. This tier also involves link marshaling [35], intra-inter domain communication, scheduling, packet transfer between flow sensing chips and the IoT gateway. Feature control requires a feature that accepts only suitable background data and excludes redundant data. Wide number of sensing chips retains a lot of features, but only a small subset of features is useful to produce background data. Function helps minimize irrelevant data sharing improving the data transfer rate of usable data and also decreases energy and CPU usage. The number of features can be different on the basis of program specifications and context data types.

4.4.3 Abstraction Tiers

One of the most critical aspects of Open-Flow is the inclusion of simulated tiers with predetermined tiers, leaving the same architecture untouched. A virtual equate can be formed between various webnet and a common platform can be built for different communication processing engines. The framework is completely centralized at the physical tiers perspective, but the delivery of the service could be retained. One central system [33] can track and control all kinds of traffic. It will help to achieve greater latency, reliability, efficient routing, etc., which can lead to better quality of amenity. Packets are exchanged through certain neighboring junction in a multi-hopping situation. As a result, junction near access points carries too much load relative to distant junction in a downstream scenario, and the inactivity of this critical junction can cause the webnet to crash. Remote presence of sensing chips junction can solve the obstacle by establishing a virtual equate ion between two sensing chips webnet by access point negotiation. So we can build a three-stage platform where a common platform and a virtualization tiers are newly introduced to the infrastructure [22]. Sensing chips should not care about their ability to enter or location except in rough areas. Packet will be sent to any junction, even though it is based on separate webnet.

4.4.4 Service Tiers

Storage governance has the concept of all kinds of unfamiliar and/or relevant technology and accomplishment that can make the gadget flexible and effective. It is not only responsible for storing records, but also for providing protection along with it. It also provides easy access to data; data aggregation to improve service orientation, and most significantly, to maximize storage availability. Storage and control tiers cover data storage and gadget control, database [46] amenity, and market strategy & processes. Although they are used in single tiers, the business support structure is directly above the cloud storage service, while Open-Flow is positioned below it as presented to provide virtualizations and display control. Service governance blends the resources needed with operational solutions, making it easy for new generation of user amenity. These upcoming resources need to be interlinked and combined in order to satisfy demand for socio-economic [44] considerations such as framework analysis, protection measurement, climate control, agricultural modernization, etc.

5 Conclusion and Future Scope

Technological enabled colonies are turning into a reality as a component of our regular daily existences. The centralized System Must Be Capable to maintain proper communication among smart channels like traffic designs, surveillance cams and Water control systems controlled by smart applications. Any security flaws can create the severe vulnerability into the system and block the entire process. Network fail due to a digital access can cause illegitimate loss of facts, property and innate danger to lives because of uncontrolled environmental conditions. With each obstacle anyway there is always a solution. We have also talked about surveillance alternatives to help ameliorate and solve a portion of the obstacle that are associated with automated metropolis surveillance. The idea of technological enabled colonies is changing the globe. While we do not realize such countless complex technologies have already coordinated with the metropolis foundation and there are various advantages associated with this idea, surveillance is significant issue thinking about the vulnerabilities and several frail links. Lot of work should be done so the residents can completely realize the advantages of the automated metropolis. In this Chapter, we have examined several surveillance perils associated with the automated metropolis through situations. We have also talked about the characteristic solutions and the inference to make sure about the automated metropolis from a cybernated surveillance view of design.

References

1. Shukla, S. K., Rawat, R., & Murthy, C. (2012). *Web Attacking Parameters Filtration: A Approach For Attack Signature Verification.*
2. Dhariwal, S., Patearia, N., & Rawat, R. (2011). C-Queued Technique against SQL injection attack. *International Journal of Advanced Research in Computer Science, 2*(5). https://doi.org/10.26483/ijarcs.v2i5.737
3. Rawat, R., Patearia, N., & Dhariwal, S. (2011). Key generator based secured system against SQL-injection attack. *International Journal of Advanced Research in Computer Science, 2*(5).
4. Nayyar, A., Puri, V., & Le, D. N. (2017). Internet of nano things (IoNT): Next evolutionary step in nanotechnology. *Nanoscience and Nanotechnology, 7*(1), 4–8.
5. Turing, A. M. (2009). Computing machinery and intelligence BT–parsing the Turing test. In R. Epstein, G. Roberts, & G. Beber (Eds.), *Philosophical and methodological issues in the quest for the thinking computer.* Springer.
6. Zielinski, D. (2017). Get intelligent on AI: Artificial intelligence can boost HR analytics, but know what you're buying. *HR Magazine, 62*(9), 60–61.
7. Solanki, A., & Nayyar, A. (2019). Green internet of things (G-IoT): ICT technologies, principles, applications, projects, and challenges. In *Handbook of Research on Big Data and the IoT* (pp. 379–405). IGI Global.
8. Anavangot, V., Menon, V. G., & Nayyar, A. (2018, November). Distributed big data analytics in the internet of signals. In *2018 International Conference on System Modeling & Advancement in Research Trends (SMART)* (pp. 73–77). IEEE.
9. Krishnamurthi, R., Nayyar, A., & Solanki, A. (2019). Innovation opportunities through internet of things (IoT) for Smart cities. In *Green and Smart Technologies for Smart Cities* (pp. 261–292). CRC Press.
10. Sen, M., Dutt, A., Agarwal, S., & Nath, A. (2013, April). Issues of privacy and security in the role of software in smart cities. In *2013 International Conference on Communication Systems and Network Technologies* (pp. 518–523). IEEE.
11. Kumar, S. A., Vealey, T., & Srivastava, H. (2016, January). Security in internet of things: Challenges, solutions and future directions. In *2016 49th Hawaii International Conference on System Sciences (HICSS)* (pp. 5772–5781). IEEE.
12. Tranfield, D., Denyer, D., & Smart, P. (2003). Towards a methodology for developing evidence-informed management knowledge by means of systematic review. *British Journal of Management, 14*(3), 207–222.
13. Valluripally, S., Sukheja, D., Ohri, K., & Singh, S. K. (2019, May). IoT based Smart luggage monitor alarm system. In *International conference on internet of things and connected technologies* (pp. 294–302). Springer.
14. Hummon, N. P., & Dereian, P. (1989). Connectivity in a citation network: The development of DNA theory. *Social Networks, 11*(1), 39–63.
15. Rawat, R., & Shrivastav, S. K. (2012). SQL injection attack detection using SVM. *International Journal of Computer Applications, 42*(13), 1–4.
16. Kitchenham, B. (2004). Procedures for performing systematic reviews. *Keele, UK, Keele University, 33*(2004), 1–26.
17. Nayyar, A. N. A. N. D., Rameshwar, R. U. D. R. A., & Solanki, A. R. U. N. (2020). Internet of things (IoT) and the digital business environment: A standpoint inclusive cyber space, Cyber Crimes, and Cybersecurity.
18. Nascimento, A. M., da Cunha, M. A. V. C., de Souza Meirelles, F., ScornavaccaJr, E., & de Melo, V. V. (2018). *A literature analysis of research on artificial intelligence in management information system (MIS).* In AMCIS.
19. Albino, V., Berardi, U., & Dangelico, R. M. (2015). Smart cities: Definitions, dimensions, performance, and initiatives. *Journal of Urban Technology, 22*(1), 3–21.

20. De Nooy, W., Mrvar, A., & Batagelj, V. (2018). *Exploratory social network analysis with Pajek: Revised and expanded edition for updated software* (Vol. 46). Cambridge University Press.
21. Ben Rjab, A., & Mellouli, S. (2019, April). *Artificial intelligence in smart cities: Systematic literature network analysis*. In Proceedings of the 12th International Conference on Theory and Practice of Electronic Governance (pp. 259–269).
22. Morri, N., Hadouaj, S., & Said, L. B. (2015, June). *Multi-agent optimization model for multi-criteria regulation of multi-modal public transport*. In 2015 World Congress on Information Technology and Computer Applications (WCITCA) (pp. 1–6). IEEE.
23. Hawi, R., Okeyo, G., & Kimwele, M. (2015). Techniques for smart traffic control: An in-depth. *International Journal of Computer Applications Technology and Research, 4*(7), 566–573.
24. Dias, G. M., Bellalta, B., & Oechsner, S. (2015, November). *Predicting occupancy trends in Barcelona's bicycle service stations using open data*. In 2015 sai intelligent systems conference (intellisys) (pp. 439–445). IEEE.
25. Pramanik, P. K. D., Solanki, A., Debnath, A., Nayyar, A., El-Sappagh, S., & Kwak, K. S. (2020). Advancing modern healthcare with nanotechnology, nanobiosensors, and internet of nano things: Taxonomies, applications, architecture, and challenges. *IEEE Access, 8*, 65230–65266.
26. Ben Rjab, A., & Mellouli, S. (2019, April). *Artificial intelligence in smart cities: Systematic literature network analysis*. In Proceedings of the 12th International Conference on Theory and Practice of Electronic Governance (pp. 259–269).
27. Kouziokas, G. N. (2017). The application of artificial intelligence in public administration for forecasting high crime risk transportation areas in urban environment. *Transportation Research Procedia, 24*, 467–473.
28. Ullah, F., Al-Turjman, F., & Nayyar, A. (2020). IoT-based green city architecture using secured and sustainable android services. *Environmental Technology & Innovation, 20*, 101091.
29. Liebig, T., Piatkowski, N., Bockermann, C., & Morik, K. (2014, March). *Predictive trip planning-Smart routing in Smart cities*. In EDBT/ICDT Workshops (pp. 331–338).
30. Ben Rjab, A., & Mellouli, S. (2019, April). Artificial intelligence in smart cities: Systematic literature network analysis. In Proceedings of the 12th International Conference on Theory and Practice of Electronic Governance (pp. 259–269).
31. Cochez, M., Periaux, J., Terziyan, V., & Tuovinen, T. (2015, May). Agile deep learning UAVs operating in Smart spaces: Collective intelligence versus "Mission-impossible". In *European Congress on Computational Methods in Applied Sciences and Engineering* (pp. 31–53). Springer.
32. Teodorovic, D. (2003). Transport modeling by multi-agent systems: A swarm intelligence approach. *Transportation Planning and Technology, 26*(4), 289–312.
33. Menouar, H., Guvenc, I., Akkaya, K., Uluagac, A. S., Kadri, A., & Tuncer, A. (2017). UAV-enabled intelligent transportation systems for the smart city: Applications and challenges. *IEEE Communications Magazine, 55*(3), 22–28.
34. McArthur, D., Lewis, M., & Bishary, M. (2005). The roles of artificial intelligence in education: Current progress and future prospects. *Journal of Educational Technology, 1*(4), 42–80.
35. Srivastava, S., Bisht, A., & Narayan, N. (2017, January). Safety and security in smart cities using artificial intelligence—A review. In *2017 7th International Conference on Cloud Computing, Data Science & Engineering-Confluence* (pp. 130–133). IEEE.
36. Bothma, R. (2017). Chatbots are coming: Technology upgrade-word of mouse. *HR Future, 2017*(Dec 2017), 36–37.
37. Yano, K. (2017). How artificial intelligence will change HR. *People & Strategy, 40*(3), 42–47.
38. Sennan, S., Ramasubbareddy, S., Luhach, A. K., Nayyar, A., & Qureshi, B. (2020). CT-RPL: Cluster tree based routing Protocol to maximize the lifetime of internet of things. *Sensors, 20*(20), 5858.
39. Kumar, A., Sangwan, S. R., & Nayyar, A. (2020). Multimedia social big data: Mining. In *Multimedia big data computing for IoT applications* (pp. 289–321). Springer.

40. Stevenson, B. (2018). Artificial intelligence, income, employment, and meaning. In *The Economics of Artificial Intelligence: An Agenda* (pp. 189–195). University of Chicago Press.
41. Shortliffe, E. H. (1993). The adolescence of AI in medicine: Will the field come of age in the '90s? *Artificial Intelligence in Medicine, 5*(2), 93–106.
42. Balaji, B. S., Raja, P. V., Nayyar, A., Sanjeevikumar, P., & Pandiyan, S. (2020). Enhancement of security and handling the inconspicuousness in IoT using a simple size extensible blockchain. *Energies, 13*(7), 1795.
43. Tarassenko, L., & Watkinson, P. (2018). Artificial intelligence in health care: Enabling informed care. *The Lancet, 391*(10127), 1260.
44. Hanson, C. W., III, & Marshall, B. E. (2001). Artificial intelligence applications in the intensive care unit. *Critical Care Medicine, 29*(2), 427–435.
45. Elmaghraby, A. S., & Losavio, M. M. (2014). Cyber security challenges in Smart cities: Safety, security and privacy. *Journal of Advanced Research, 5*(4), 491–497.
46. Bellazzi, R., & Zupan, B. (2008). Predictive data mining in clinical medicine: Current issues and guidelines. *International Journal of Medical Informatics, 77*(2), 81–97.
47. Rawat, R., Dangi, C. S., & Patil, J. (2011). Safe guard anomalies against SQL injection attacks. *International Journal of Computer Applications, 22*(2), 11–14.
48. Rawat, R., & Zodape, M. (2012). URLAD (URL attack detection)-using SVM. *International Journal of Advanced Research in Computer Science and Software Engineering, 2*(1).

Role of Artificial Intelligence and IoT in Next Generation Education System

Kiran Ahuja and Indu Bala

1 Introduction

Before the introduction of computers and other digital technologies in classrooms, traditional face to face classroom teaching was in the trend where the knowledge was purely disseminated solely employing human effort. The introduction of personal computers and the latest developments in computers, the internet, software technology, and fast computational capabilities have led to the adoption of computers in almost every sector including the education sector [1].

Computer and information technologies have evolved continuously to develop Artificial Intelligence (AI). Coppin has to define Artificial Intelligence as *"The ability of machines to adapt to new situations, deal with emerging situations, solve problems, answer questions, device plans, and perform various other functions that require some level of intelligence typically evident in human beings"* [2]. Another popular definition of AI by Whitby is *"The study of intelligent behavior in human beings, animals, and machines and endeavoring to engineer such behavior into an artifact, such as computers and computer-related technologies"* [3]. Similar to other technologies, AI has also penetrated deeply into the education sector in various capacities. The use of AI in the education sector has dramatically improved its efficiency in global and personalized learning and smarter content creation [4]. It is still evolving and new developments are coming into the market now and then for school administration, effective teaching, and better learning. The use of AI-enabled applications/systems in the education sector is gaining momentum every year.

K. Ahuja (✉)
DAV Institute of Engineering and Technology, Jalandhar, India

I. Bala
Lovely Professional University, Phagwara, India

© The Author(s), under exclusive license to Springer Nature Switzerland AG 2021
F. Al-Turjman et al. (eds.), *Intelligence of Things: AI-IoT Based Critical-Applications and Innovations*,
https://doi.org/10.1007/978-3-030-82800-4_8

AI has the potential to transform traditional classroom teaching and content delivery strategies. The AI-based educational technologies such as virtual and augmented reality can deliver a real-time classroom learning experience to students. The dependence on automated ERP management systems by educational institutions for various administrative operations like course registration, course material sharing, staff recruitments, and appraisals, hostel management, online fee collection, and other day-to-day activities about academic operations reduces human involvement [5].

For instance, a conventional student information management system (SIS) needs humans to register a student in a particular course based on his/her academic performance such as rank, scores in previous semesters, etc. The operator assigns the relevant course to the learner from available options. The whole process reduces the management's efforts of analyzing and processing individual student's applications; however, it still requires human endeavors to perform academic operations [6]. Now, imagine a scenario in which the whole process is done using an AI-driven system, right from automatically receiving an online student application form, assigning the right course to the candidate based on his preferences filled and too without any human interference. Since, data security is big concerns these days for educational institutions as their databases have a lot of students sensitive information related to their admissions, registrations, attendance, assessments, grades, reappears, awards and recognitions, etc., the use of an AI-enabled system eliminates the risk of data loss and cyber hacks [7].

The one section of individuals, who benefit extremely by adopting AI technology, is the academic management personnel (teachers/faculty/teaching staff.) The AI's ability to auto-grade papers, auto-evaluate assignments, assign homework, and many more makes it a perfect tool for educators to simplify their job enormously. The capacity of AI does not just end there, It enhances personalized and individualized learning by seamlessly integrating with LMS to share study materials, navigate through lesson content, create and view educational videos, presentations, illustration artwork, images, audio lessons, etc. And, at the same time, it allows students to attend tests and exams online on smartphones, tablets, computers, and various other electronic devices [8].

Moreover, AI innovation encourages e-learning, to enhance personalized, blended, competency-based, and differentiated learning methods by making, utilizing, and overseeing proper technological procedures and assets in teaching. Additionally, AI in education incorporates different frameworks and tools that emphasize structure, improvement, research and development, administration, and assessment of procedures to ease teaching and learning [9].

The rest of the chapter is organized as follows: Section 2 describes the benefits of AI in the education system. The importance of AI in the new education system is described in Sect. 3. Section 4 provides the glimpses of AI-based educational solutions in trend. However, Sect. 5 provides information about AI-enabled online learning platforms. In Sect. 6, IoT applications in education are explained in detail. In Sect. 7 potential of AI and IoT are discussed followed by the conclusion in Sect. 8.

2 Boons of AI in the Education System

Indeed, AI has revolutionized the educational industry. There are many benefits associated with the usage of Artificial Intelligence in educational institutions not only for the students but also for the teachers. In this section, we have discussed some of them like and pictorial view is provided in Fig. 8.1:

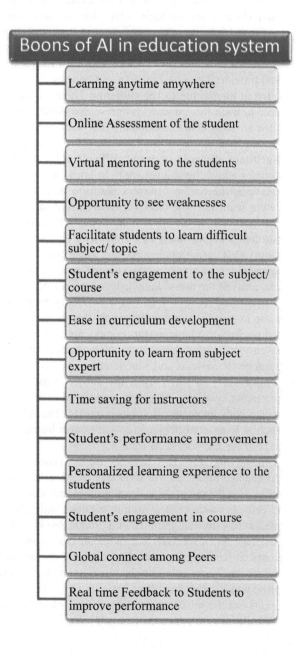

Fig. 8.1 Pictorial view of benefits of AI in the education system

- **Learning anytime anywhere:** Youngster spends a lot of time to manage their day to chores using smartphones, laptops, or tablets. AI-based online learning platforms facilitate them to manage their study hours according to their convenience [10]. Moreover, they can also get feedback and help from their online course instructors in a real-time mode.
- **Online assessment of the student:** AI-based solutions are capable to assess the student's intellectual level and his interest in topics. Such a system facilitates students to work on their weak sides in academics and offers/suggest them additional learning materials based on their weaknesses [11, 12]. For example, an AI-enabled app can be asked to the student to appear in the test before using it. Based on the test performance, the app analyses the learner's intellectual level and provides him suitable tasks and courses.
- **Virtual mentoring to the students:** There are many AI-based platforms available to track the students' progress. No doubt, human teachers can understand their students better emotionally, but such systems are good enough to provide instant feedback to the teachers and act as a virtual tutor.
- **Facilitate students to learn difficult subjects/topics:** Many online learning courses allow course instructors to know the gaps in students' knowledge. For example, the online Coursera platform notifies the course instructor if many online learners chose incorrect answers to a particular question [13]. As a result, the instructor has an opportunity to pay attention to the demanded topic.
- **Student's engagement to the subject/ course:** Latest technologies like virtual reality (VR) and gamification, etc. make learning contents more interactive and keep students engaged in the subject/ course [14]. Many AI-enabled algorithms can analyze the users' knowledge/ performance and interests and provide more personalized recommendations and training programs.
- **Ease in curriculum development:** Recent developments in AI have benefitted the teachers to a large extent. For example, curriculum development has become so easy to develop these days which otherwise requires lots of time in searching appropriate contents and educational materials to be included [15].
- **Opportunity to learn from the subject expert:** There are many educational platforms available online and students always have an opportunity to communicate with international subject experts. The AI-enabled educational platform such as Coursera offers appropriate course instructors based on his/her teaching experience and soft skills [16].
- **Time-saving for instructors:** Some people have bizarre thoughts that AI-enabled machines will soon take over teacher's jobs. On contrary to this, the fact is that the use of AI in the education sector enhances teacher's efficiency and effectively manages their routine teaching, and yields good outcomes. For example, an AI-powered grammar and pronunciation tool can refine learners' English language skills. Whereas in a traditional classroom teaching a teacher spends long hours correcting grammatical mistakes for all the students in a class. Moreover, the teacher is just able to point out the grammatical error to the students without providing a detailed explanation of it. On contrary to this, AI-powered software points out the mistakes, provides a detailed explanation of them, and also provides illustrations of correct usage [17].

Further, AI-based devices can be efficiently designed for teachers to save their time by doing some of the routine tasks like attendance marking, presenting study material on connected display screens as per the instructional schedule, and helping teachers navigate through the displayed content with ease. AI-enabled educational apps can gauge each student's understanding of a concept individually and accurately in the shortest period possible. Such apps are more learner-centric and offer personalized evaluation as compared to the regular test given to the students with similar questions [18]. AI apps can assist the teaching-learning processes at a much lower cost and faster pace as compared to human mediators.

- **Student's performance improvement:** Artificial intelligence is becoming a part of our daily routine. So, we can anticipate that it will also make changes in the realm of education. Apart from teaching, teachers perform multiple tasks and student grading is one of them. Artificial Intelligence-enabled educational tools help teachers to grade their students by gathering information about an individual student. Such tools are useful in measuring individual student's strengths and weaknesses and recommend solutions too [19].

  One of the major contributions of AI in education is its ability to improve student's weaknesses in the classroom rather than to wait for the results of exams to see how students are performing. AI-enabled tools or apps allow teachers to see how students are learning throughout the lessons. Teachers will be able to monitor progress easier and more regularly. Also, AI will recognize when a lesson needs to be retaught, such as when a majority of students do not master a certain skill or miss a type of question. Artificial Intelligence and education will go hand-in-hand in the future, and it will create more opportunities for success for both teachers and students [20].

- **Personalized learning experience for the students:** AI-driven technologies, such as data analytics and machine learning, are already creating major disruptions in the education segment. Top online learning platforms are using these cutting-edge technologies to measure each user on their skill sets, capabilities, and interests before making the most contextual course recommendations [8]. This tech-led intervention enables online learning platforms to act as guides and advisors to a candidate's upskilling/reskilling journey and helps job seekers accelerate their career growth and access much better professional opportunities by gaining relevant, in-demand skills.

  This, however, is just the tip of the iceberg. AI and related technologies can play an even bigger role in both traditional and non-traditional learning environments. AI systems offer a more personalized learning experience to the students based on the individual learning speeds and needs of different students. This is something that even the best of educators often find challenging at present. AI-led technologies also find applications in coursework curation, helping educators to understand the need for inclusion of various topics for the students and to design the most appropriate curriculum for them.

- **Student's engagement in the course:** Artificial Intelligence-based educational tools/ devices have the potential to engage the student by making the learning experience more fun activity. It allows students to captivate more in a class and to

comprehend the concepts taught employing simulations and gaming technologies.
- **Global connect among peers:** The use of AI converts a normal classroom into a global classroom irrespective of the actual geographical location of the student. If students are somehow not able to attend regular class due to some reason, they still have access to the delivered lecture with just a click on a link [21]. It allows the student to experience live classroom virtually. Moreover, it also enables students to interact with their peers worldwide. Not only that, it allows students to interact with their teachers as well as other course-related experts.
- **Real-time feedback to students to improve performance:** Many times students hesitate asking questions or clearing their doubts in a class from teachers in front of their peers. They may also uneasy in getting critical advice on their performance publically. However, AI-enabled systems make students feel comfortable getting advice on their mistakes to learn the subjects well. Such systems also provide necessary suggestions to the students to improve their scores.

3 AI-Based Technologies for Education

AI has the potential to automate administrative tasks for teachers and educational institutions. Teachers spend most of their time developing course curriculum, lecture notes preparation, students grading, evaluations, etc., the AI platforms can reduce their burden significantly if implemented properly. In this section, we have discussed some of how the educational industry has revolutionized using AI. Some of the typical implementations in the education sector are represented in Fig. 8.2.

- **Personalized learning:** Artificial Intelligence offers students personalized learning experience according to their needs. Carnegie Learning, for example, is the educational platform that uses AI to provide more personalized courses to the students. AI tools understand the needs of the students and create individual instructions, tests, and feedback procedures to fill the gaps in their knowledge. Moreover, it also can scan and analyze students' facial expressions to understand their difficulties with the subject and can change the course contents and instructions according to their needs [22].

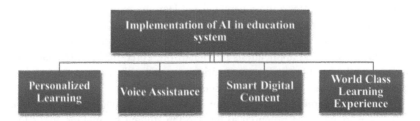

Fig. 8.2 Diverse implementation criteria of AI in the education system

- **Voice assistance:** Apple Siri, Amazon Alexa, and Google Home are some interactive voice assistance services that can be used for learning by students anytime and anywhere without a teacher. Such assistance is already been used by Arizona State University students to follow their scheduled activities.
- **Smart digital content:** Smart digital content represents the online instructional materials from computerized textbooks to personalized interfaces. The content development companies like Content Technologies, Inc. use AI to automate their business processes and to provide users better solutions for the education field [23]. Similarly, Cram101 is helpful to the learners by breaking the textbook's content into parts, includes a chapter summary, tests, and so on.
- **World-class learning experience:** AI-based interactive online apps facilitate students to share knowledge worldwide. There are many online interactive platforms where a student can learn courses from the world's best teachers/ techies. In addition to this, AI also facilitates providing opportunities for the students to interact with each other despite the language barriers by adding subtitles to the learning material. AI-based solutions like presentation translator can be extremely helpful for the students who have hearing or visual problems by creating subtitles in real-time mode or by using speech recognition for the students to hear in their native languages, respectively.

4 AI-Based Educational Solutions: Current Scenario

In this section, some of the educational applications that use AI to improve learning in students of all ages are discussed. These applications empower both learners and teachers with more avenues for reaching their educational goals. The major solutions applied in the current situation are shown in Fig. 8.3.

5 Development of AI-Enabled Online Learning Platforms

In this section, a stepwise procedure to build an e-learning website using Artificial Intelligence is discussed. To create such contents six main steps are as follows and represented in Fig. 8.4:

- **Step 1**—Know your Competitors: Before embarking upon the solution, the little survey is a must to know your competitors in the market. Customers always demand interactive, easy to use applications/ solutions with more new features. Knowledge about the existing platforms is always helpful to add more interesting and creative concepts for your project.
- **Step 2**—*Creation of content repository*: While creating a solution for the education sector, one always needs useful and interesting content. To do so, one can collaborate with tutors from various universities, schools, or colleges.

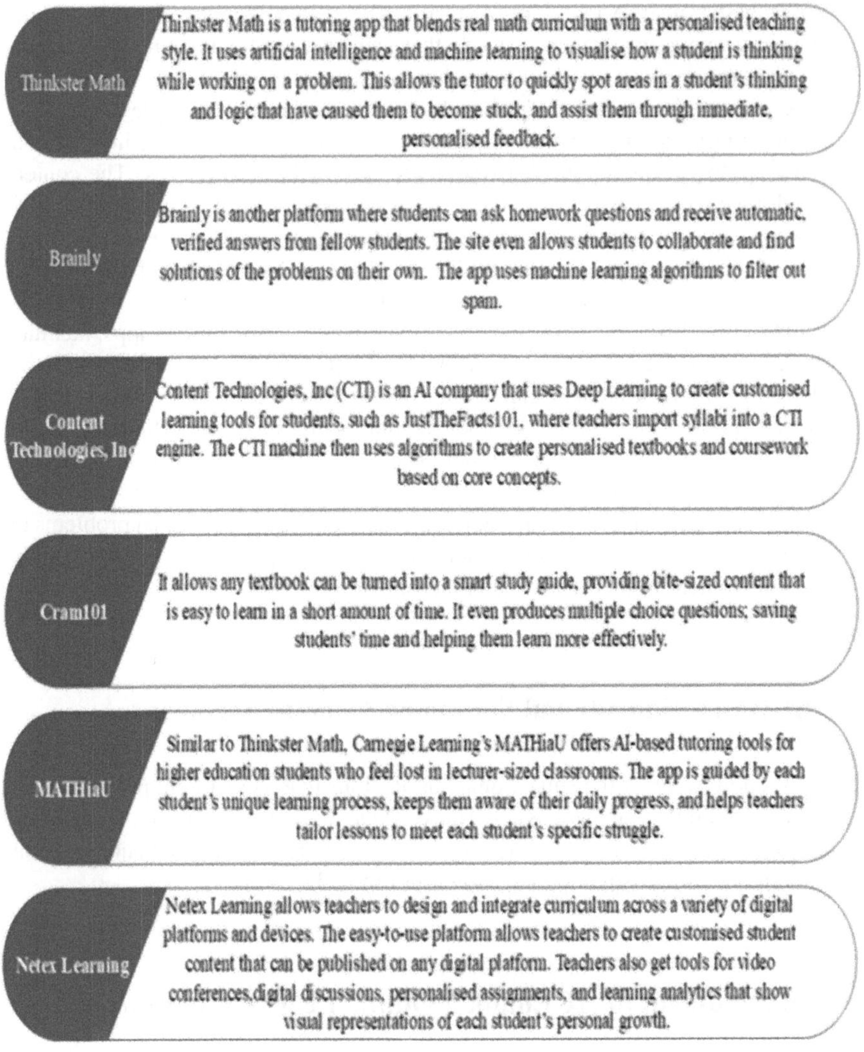

Thinkster Math — Thinkster Math is a tutoring app that blends real math curriculum with a personalised teaching style. It uses artificial intelligence and machine learning to visualise how a student is thinking while working on a problem. This allows the tutor to quickly spot areas in a student's thinking and logic that have caused them to become stuck, and assist them through immediate, personalised feedback.

Brainly — Brainly is another platform where students can ask homework questions and receive automatic, verified answers from fellow students. The site even allows students to collaborate and find solutions of the problems on their own. The app uses machine learning algorithms to filter out spam.

Content Technologies, Inc — Content Technologies, Inc (CTI) is an AI company that uses Deep Learning to create customised learning tools for students, such as JustTheFacts101, where teachers import syllabi into a CTI engine. The CTI machine then uses algorithms to create personalised textbooks and coursework based on core concepts.

Cram101 — It allows any textbook can be turned into a smart study guide, providing bite-sized content that is easy to learn in a short amount of time. It even produces multiple choice questions; saving students' time and helping them learn more effectively.

MATHiaU — Similar to Thinkster Math, Carnegie Learning's MATHiaU offers AI-based tutoring tools for higher education students who feel lost in lecturer-sized classrooms. The app is guided by each student's unique learning process, keeps them aware of their daily progress, and helps teachers tailor lessons to meet each student's specific struggle.

Netex Learning — Netex Learning allows teachers to design and integrate curriculum across a variety of digital platforms and devices. The easy-to-use platform allows teachers to create customised student content that can be published on any digital platform. Teachers also get tools for video conferences, digital discussions, personalised assignments, and learning analytics that show visual representations of each student's personal growth.

Fig. 8.3 Shows the existing AI-based solutions in the education system

Furthermore, learning materials can also be obtained from various sources like training programs and courses, etc.

- **Step 3**—*Know your Project Requirements*: Before developing some solution, one must know the project requirements and business goals. The constant feedback from users and regular updating app/solution by adding new features are helpful for the success of the project.
- **Step 4**—*Testing and debugging*: To create a huge customer base, service providers should always provide them great user experience. Thus, before launching

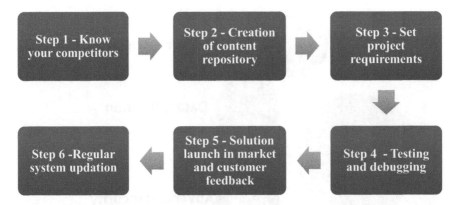

Fig. 8.4 Represents the various steps included for the development of an AI platform for the education system

solutions in the market, the AI platform must be properly tested and bugs (if any) must be fixed by qualified quality control engineers.
- **Step 5**—*Solution launch in market and customer Feedback*: As discussed previously, it is always beneficial to create a minimum viable product (MVP0) version of the solution to get customers' feedback. Their feedback can be beneficial to know the scope of improvement in the project. Additionally, teachers and learners can also communicate their expectations and requirements for your product/ solution.
- **Step 6**—*Regular System updation*: It is mandatory to update your solution/ platform regularly by adding new exciting features to remain competitive in the market. It is always better to tie up with new tutors and to launch new training programs from time to time. The integration and use of Artificial Intelligence have always had several great solutions to be offered to adults, children, tutors, and even educational institutions.

6 IoT Applications in Education

Technology can contribute significantly to improving education rather than hiring new staff in schools, colleges, and universities. IoT can help teachers as well as students by lending them great technological help to unleash their talents and to open up new avenues for them. It has the potential to enable students to comprehend the complex mathematical formulas and to understand the basic notion of the subject. It involves the students in classroom activities and the learning process as a fun develops the interest of the student in a subject by making it more interactive by managing the balance of modern and contemporary teaching methods to uplift the

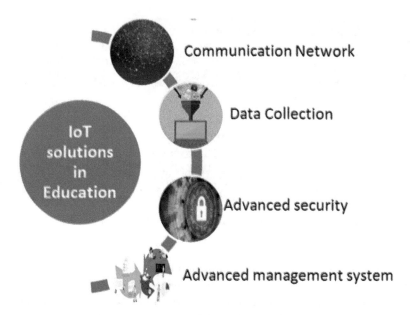

Fig. 8.5 Innovative implications of IoT in teaching

society. Technological advances have made education accessible and interactive for every student in the world [23–26]. A few innovative implications are presented in Fig. 8.5.

6.1 Smart Infrastructure

With the advancement in technologies, the teaching pedagogies are also in transition. These days' students are enjoying learning on smart interactive boards more than conventional blackboards. The test and figures can be illustrated in a better way as compare to the blackboard or textbooks to facilitate the students to understand the concept of a lesson. Thus, the introduction of IoT in education is making education more interactive and the exchange of information simple and effective manner from instructor to the students [27–33]. With the introduction of smart boards in the classrooms, the teaching job is becoming a lot easier and interesting with a better illustration of graphs, and mathematical formulas through audio-video and storage facilities. Computer-based intelligence can be the most effective method for personalization and smoothing out the administrative tasks while permitting the instructor(s) an opportunity to give comprehension and versatility to the learners. Due to these facts, smart infrastructure is further divided into three major categories as follows:

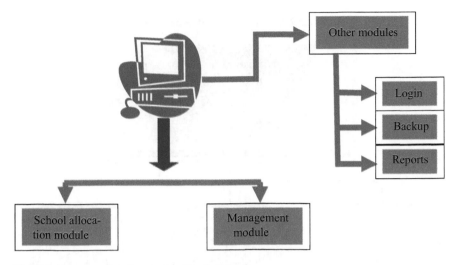

Fig. 8.6 Resource allocation at school/institute level

6.1.1 Computer Resource Allocation in Schools

A school that gets resources since it has understudies with a specific blend of adapting needs has the obligation of giving projects to address those issues, has the power to settle on choices on how those resources will be apportioned, and ought to be responsible for the utilization of those assets, remembering results for connection to adapting needs. Choices on the resource portion should possibly be made midway on the off chance that they cannot be made locally [34–36]. Choices on things of use should possibly be prohibited from the School Global Budget if schools do not control expenditure, if there is an inordinate variety of usage, if usage designs are eccentric, if usage is once-off, or use for which schools are paid courses. The process of resource allocation is represented in Fig. 8.6.

6.1.2 Multimedia Classrooms

Multimedia in the classroom could incorporate PowerPoint introductions that are made by the educator, business programming, (for example, sight and sound reference books) that is utilized for reference or guidance, or exercises that straightforwardly connect with the students in utilizing multimedia to build and pass on information. For the motivations, we mainly focused on drawing interest in students for the utilization of the multimedia to develop and convey the information. Instances of media, at that point, could include: Students utilizing idea planning programming, (for example, Inspiration) to conceptualize [37–42]. Students utilizing a bookkeeping page or diagramming mini-computer to record information and produce graphs. A little gathering of students is making a computerized film to show a technique. A class website shows understudy work of art students by checking their hands and also bringing their pictures into PowerPoint for an introduction about

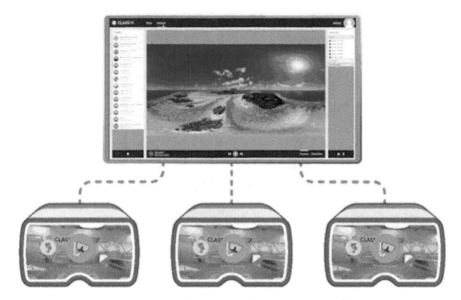

Fig. 8.7 Virtual reality for the classroom via multimedia implementation

fingerprints. Figure 8.7 is included to exemplify the usage of multimedia in the latest style classrooms.

6.1.3 Digital Resource for Schools

Innovation will keep on changing how instructors educate and how students learn data. While numerous advanced resources can assist instructors with getting ready students for the difficulties ahead, it is dependent upon teachers to assist students with gaining the outlook and abilities important to prevail in the intricate world. The challenge for instructors turns out to be the way when they have to actualize computerized resources in classrooms to engage students for effective learning. Probably the simplest approaches to incorporate a computerized resource are to discover places in a lesson plan where an idea or substance can be fortified. Instructors are continually searching for approaches to rejuvenate an exercise. Before re-composing your next exercise plan, look at user-friendly digital resources [43] as shown in Fig. 8.8, which could properly challenge students' attitudes and move development in learners.

6.2 Students Attendance Management

Many education institutes believe that students must have some significant percentage of class attendance to appear in the examination. With IoT, the attendance of the

Fig. 8.8 User-friendly digital resources for classrooms

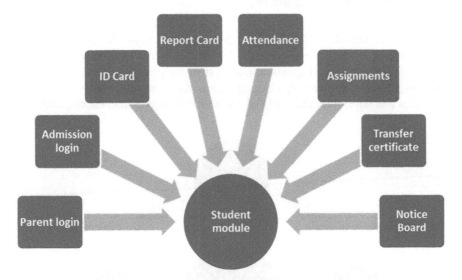

Fig. 8.9 A pictorial view of student module facilities

students can be managed accurately without human error. Using location coordinates of the student's hostel in real-time, it is easy to trace the student activity on campus. With an efficient IoT based attendance system, student's attendance calculation, regularity and punctuality check becomes effortless and less time-consuming. The regular attendance feedback to the students and parents makes students more regular in their classes. The various options related to student performance evaluation are shown in Fig. 8.9.

6.3 Safety Measures to the Institutional Infrastructure

Institutions are investing in their infrastructures by deploying more IoT sensors such as short circuit protection, fire alarm, Wi-Fi clocks, video monitoring, earthquake, and serious weather changes, etc. to avoid any kind of danger of natural calamity. On the activation of any such alarms, the IoT sensors instantly detect and send the alert message to reverse the situation. Figure 8.10 shows some of the IoT based smart safety measures.

6.4 Technological Advancement as a Learning Privilege to the Disabled Students

Earlier, it was unthinkable for disabled kids to learn the subject. With technological advancements, it has become a lot easier for such students to learn new things. The innovative designs of hearing aids, a smart system comprised of smart gloves with a tablet to generate audio from sign language, audio to text writer are some noteworthy examples that have made the world of disabled kids brighter and earning easy and

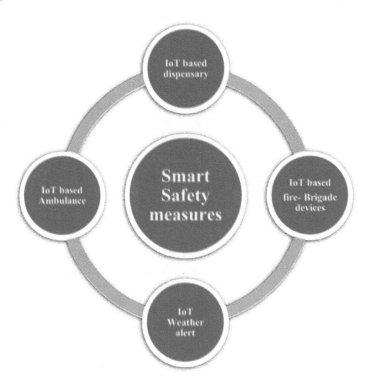

Fig. 8.10 Smart safety measures at school/institutional level

Fig. 8.11 IoT based assistance for disabled students

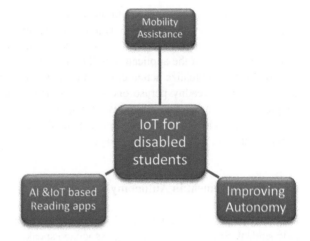

interesting [44]. When IoT is utilized properly, then it opens up innovative prospects for disabled persons by providing the availability of connected technology. A few applications in this regard are presented in Fig. 8.11.

6.4.1 Mobility Assistance

A couple of IoT-enabled devices help individuals with handicaps beat mobility issues. For example, think about Crosswalk, an assistive application situated in the Netherlands. It is downloadable on cell phones and it permits crippled people on foot to alarm the traffic signals and solicitation additional time for crossing over. A client may look over four sorts of settings as per their mobility. This IoT item shows amazing interoperability all the while cooperating with GPS and the software of traffic signals. Improvements like these guide in separating transportation obstructions that are looked at by incapacitated people, permitting them to arrive at public vehicle center points and make mobility more comfortable. Furthermore, IoT-enabled advancements like this permit incapacitated people to briefly conquer the current absence of versatile and assistive foundation and strategy by giving people a practical option [45, 46].

6.4.2 Reading Surroundings

IoT and AI joined together can help handicapped people better peruse and comprehend their environmental factors. This is particularly helpful for individuals who are outwardly disabled. Considering Microsoft's Vision AI application which is intended for outwardly moving individuals to find out about their nearby environmental factors. The man-made reasoning customized application can coordinate walkers from swarmed intersections and can even illuminate individuals about what others around them are communicating facially. A comparative innovation called Cloud Vision API has been acquainted by Google for engineers with make applications and gadgets that have acknowledgment and grouping highlights. The counterfeit "sight" of the application will extraordinarily assist clients with comprehension and contextualize actual circumstances in their environmental factors. The capacity to more readily peruse one's environmental factors can assist debilitated people with defeating the numerous correspondence hindrances they face. These improvements advance better arrangement and can be very useful to those with visual, hear-able, and intellectual incapacities.

6.4.3 Improvement in Autonomy

IoT as a smart gadget is very useful in assisting impaired people with becoming independent. Smart homes including associated speakers, coolers, stoves, and indoor regulators assist impaired people with having control of their day by day activities and can be customized to suit a particular client's requirements. IoT innovation additionally accommodates wearable-like smart-watches that interpret content including emails and messages into Braille or read the equivalent out loud. By improving self-rule, IoT is assisting impaired people with defeating both social boundaries and attitudinal obstructions and permitting them to improve their overall personal satisfaction through the arrangement of available and valuable hardware. Imaginative new improvements in this field incorporate smart insoles by Ducere Technologies, an insertable insole that utilizes vibrations to ease route. Different firms are likewise trying different things with new arrangements, for example, wearable gadgets that work like radar, and bone-conducting innovation to help route. These advancements are useful in beating actual obstructions that disabled people face, for example, steps that may keep somebody from entering stairs, or checks that could somehow keep a debilitated individual from utilizing the walkway.

7 Potential Use of AI and IoT in Education Sector

Emerging technologies like Artificial Intelligence (AI), Machine Learning (ML), and the Internet of Things (IoT) have transformed the way industries used to operate [19, 47, 48]. If we talk about the education sector, the way tutoring was delivered to the learner is completely changed. Those traditional teaching methods are getting phased out in this connected world.

The impact of AI and IoT in the education sector is unlike any other sector, unique and the link between digital technology and education is unique and appreciative. A smart school (a school that uses IoT), with the facilities operating smoothly, promotes a higher level of personalized learning. The institutional campuses use smart devices and utilize Wi-Fi network for receiving instructions and sending data. The major proponent which makes the Internet of Things and AI a very attractive technological inclusion for education is the fact that it promotes real-time learning. With the help of this and other actionable insights, lecturers can monitor student's performance on the go. Students will then be encouraged to bring their wireless devices to the classroom, as opposed to getting reprimanded for it.

Operational efficiency will be greatly improved with the help of these technologies. In an environment where several devices are connected, educators can work on creating a more dynamic classroom experience. In a situation where attendance is supposed to be recorded, this will be made incredibly simple by giving students wearable devices that can track ECG patterns. There is also the possibility for EEG sensors to be used for monitoring the cognitive abilities of students while they are attending the lecture. With all of these potentials for the Internet of Things and Artificial Intelligence in the educational sphere, the possibilities for learning are endless.

8 Conclusion

The basic notion of this chapter was to study the impact of AIIoT on education. The introduction of computers and related technologies has accelerated the use of AI in the education sector. AI in education entered in the form of computers, and later, the era of web-based education has started evolving. The use of various online platforms and learning tools has improved teacher effectiveness and efficiency of many folds. Similarly, AI has provided students better learning experiences because of the customization and personalization of learning materials availability to cater to their needs and capabilities of students. Overall, AI has touched almost all nooks and corners of the education sector, especially in administration, teaching, and learning aspects or within the context of individual learning institutions.

8.1 Future Scope

The traditional learning approaches in use to date are limited to the geographical location only. However, the introduction of the online learning teaching methodologies and Massive Open Online Courses (MOOCs) has opened up new horizons to the students. It is anticipated that AI-enabled learning platforms will increase by 47.5% by 2021. The experts from the education industry believe that the critical presence of teachers is irreplaceable in the education system, however; the conventional education system needs a revolutionary technological intervention. Thus, the teachers must be ready to adopt the best educational practices to grasp the change as the AI-driven educational platforms will facilitate the students to learn the basic notion of the subject(s) with higher accuracy. With AI engines, students can focus on strengthening their understanding of the subject(s) of interest. AI-enabled education will utilize machine intelligence and human endeavors to provide a strong foundation to the students and to make them ready for the future, where AI is the reality. The AI-based education system has opened up the doors of personalized education anytime, anywhere for the students, which was unthinkable earlier for students as well as for educators.

References

1. Flamm, K. (1988). *Creating the computer: Government, industry, and high technology.* Brookings Institution Press.
2. Coppin, B. (2004). *Artificial Intelligence Illuminated.* Jones and Bartlett.
3. Whitby, B. (2008). *Artificial intelligence: A beginner's guide.* Oneworld.
4. Timms, M. J. (Jan. 2016). Letting artificial intelligence in education out of the box: Educational cobots and smart classrooms. *International Journal of Artificial Intelligence Education, 26*(2), 701–712.
5. Snyder, H. (Nov. 2019). Literature review as a research methodology: An overview and guidelines. *Journal of Business Research, 104*, 333–339.
6. Sharma, R. C., Kawachi, P., & Bozkurt, A. (2019). The landscape of artificial intelligence in open, online and distance education: Promises and concerns. *Asian Journal of Distance Education, 14*(2), 1–2.
7. Wartman, S. A., & Combs, C. D. (Aug. 2018). Medical education must move from the information age to the age of artificial intelligence. *Academic Medicine, 93*(8), 1107–1109.
8. Roll, I., & Wylie, R. (Feb. 2016). Evolution and revolution in artificial intelligence in education. *International Journal of Artificial Intelligence Education, 26*(2), 582–599.
9. Sutton, H. (Jan. 2019). Minimize online cheating through proctoring, consequences. *Recruiting Retaining Adult Learners, 21*(5), 1–5.
10. Ahuja, K., & Bala, I. COVID-19: Creating a paradigm shift in Indian education system, Ch. 10. In *Emerging technologies for battling Covid-19: Applications and innovations* (p. 2021). Springer. Inpress.
11. Chang, C. W., Lee, J. H., Chao, P. Y., Wang, C. Y., & Chen, G. D. (2015). Exploring the possibility of using humanoid robots as instructional tools for teaching a second language in primary school. *Journal of Educational Technology & Society, 13*(2), 13–24.
12. Jones, A., Bull, S., & Castellano, G. (2018). I know that now, I'm going to learn this next' promoting self-regulated learning with a robotic tutor. *International Journal of Social Robotics, 10*(4), 439–454.
13. Nunn, S., Avella, J. T., Kanai, T., & Kebritchi, M. (Jan. 2016). Learning analytics methods, benefits, and challenges in higher education: A systematic literature review. *Online Learning, 20*(2), 1–17.
14. Estevez, J., Garate, G., & Graña, M. (2019). Gentle introduction to artificial intelligence for high-school students using scratch. *IEEE Access, 7*, 179027–179036.
15. Kim, Y., Soyata, T., & Behnagh, R. F. (2018). Towards emotionally aware AI smart classroom: Current issues and directions for engineering and education. *IEEE Access, 6*, 5308–5331.
16. Global Development of AI-Based Education, Deloitte Res., Deloitte China, Deloitte Company, 2019.
17. Lin, P.-H., Wooders, A., Wang, J. T.-Y., & Yuan, W. M. (Sep. 2018). Artificial intelligence, the missing piece of online education? *IEEE Engineering Management Review, 46*(3), 25–28.
18. Le, N. T., Strickroth, S., Gross, S., & Pinkwart, N. (2013). A review of AI supported tutoring approaches for learning programming. In *Advanced computational methods for knowledge engineering*. Springer.
19. Weiguo, W. U. (2015). Research progress of humanoid robots for mobile operation and artificial intelligence. *Journal of Harbin Institute of Technology, 47*(7), 1–19.

20. Crowe, D., LaPierre, M., & Kebritchi, M. (Jul. 2017). Knowledge based artificial augmentation intelligence technology: Next step in academic instructional tools for distance learning. *TechTrends, 61*(5), 494–506.
21. Peredo, R., Canales, A., Menchaca, A., & Peredo, I. (Nov. 2011). Intelligent Web based education system for adaptive learning. *Expert Systems with Applications, 38*(12), 14690–14702.
22. Mikropoulos, T. A., & Natsis, A. (Apr. 2011). Educational virtual environments: A ten-year review of empirical research (1999–2009). *Computers in Education, 56*(3), 769–780.
23. Chang, C. W., Lee, J. H., Chao, P. Y., Wang, C. Y., & Chen, G. D. (2015). Exploring the possibility of using humanoid robots as instructional tools for teaching a second language in primary school. *Journal of Educational Technology & Society, 13*(2), 13–24.
24. Chen, S., et al. (2014). A vision of IoT: Applications, challenges, and opportunities with China perspective. *IEEE Internet of Things Journal, 1*(4), 349–359.
25. JuniperResearch, Internet of Things' Connected Devices to Almost Triple to over 38 Billion Units by 2020 (2015).
26. Agarwal, S., & Pati, S. (2016). Study of internet of things. *International Journal for Scientific Research & Development, 4*(05), 4.
27. Aldowah, H., Ghazal, S., & Muniandy, B. (2015). Issues and challenges of using E-learning in a Yemeni Public University. *Indian Journal of Science and Technology, 8*(32), 1–19.
28. Aldowah, H., Rehman, S. U., Ghazal, S., & Umar, I. N. (2017). Internet of Things in higher education: a study on future learning. *Journal of Physics: Conference Series, 892*(1), 012017.
29. Sánchez-Torres, B., Rodríguez-Rodríguez, J. A., Rico-Bautista, D. W., & Guerrero, C. D. (2018). Smart campus: Trends in cybersecurity and future development. *Revista Facultad de Ingeniería, 27*(47), 104–112.
30. Elsaadany, A., & Soliman, M. (2017). Experimental evaluation of Internet of Things in the educational environment. *International Journal of Engineering Pedagogy., 7*, 50–60.
31. Rico-Bautista, D., Medina-Cárdenas, Y., & Guerrero, C. D. (2019). Smart University: A review from the educational and technological view of internet of things. In *International Conference on Information Technology & Systems* (pp. 427–440). Springer.
32. Bagheri, M., & Movahed, S. H. (2016). The effect of the Internet of Things (IoT) on education business model. In *2016 12th IEEE International Conference on Signal-Image Technology & Internet-Based Systems (SITIS)* (pp. 435–441).
33. https://www.biz4intellia.com/blog/iot-applications-in-education-industry/
34. Koper, R. (2014). Conditions for effective smart learning environments. *Smart Learning Environment, 1*(5).
35. Caldwell, B. J. (1997). *Principles and practices in resource allocation to schools under conditions of radical decentralization.* Paul D. Planchon, Associate Commissioner (pp. 121–136).
36. https://www.theedadvocate.org/allocating-resources-to-improve-student-learning/
37. https://fcit.usf.edu/multimedia/overview/overviewa.html
38. http://www.ascd.org/publications/educational-leadership/apr94/vol51/num07/What-Multimedia-Can-Do-in-Our-Classrooms.aspx
39. https://www.classvr.com/page/2/
40. Ghazal, S., Samsudin, Z., & Aldowah, H. (2015). Students' perception of synchronous courses using Skype-based video conferencing. *Indian Journal of Science and Technology, 8*(30).
41. Mineraud, J., et al. (2016). A gap analysis of internet-of-things platforms. *Computer Communications, 89*, 5–16.
42. Abd-Ali, R. S., Radhi, S. A., & Rasool, Z. I. (2020). A survey: The role of the internet of things in the development of education. *Indonesian Journal of Electrical Engineering and Computer Science, 19*(1), 215–221.
43. https://www.thetechedvocate.org/consider-six-digital-resources-classroom
44. https://www.kdnuggets.com/2018/04/role-iot-education.html

45. https://theiotmagazine.com/how-iot-breaks-barriers-for-people-with-disabilities-e695eaeafa15
46. Domingo, M. C. (2012). An overview of the Internet of Things for people with disabilities. *Journal of Network and Computer Applications., 35*(2), 584–596. ISSN 1084–8045.
47. https://medium.com/@soluloid/outstanding-role-of-ai-and-iot-in-education-sector-5eb4048594be
48. https://robusttechhouse.com/role-of-iot-and-ai-in-advancing-education/

Social Media Data Analysis: Rough Set Theory Based Innovative Approach

K. Anitha

1 Introduction

The purpose of Social Media Data Analysis (SMDA) is to analyse the data from Social Networks like Instagram, Twitter, Facebook, LinkedIn, Research Gate and GitHub. This process is useful for marketers to get the feedback for their products. Sponder [1] mentioned in their book that "The art and science of extracting valuable hidden insights from vast amounts of semi-structured and unstructured social media data to enable informed and insightful decision making". SMDA is essential for optimal decision making. Identification of data, analysis of data and its interpretation are necessary and sufficient steps should be followed in this process. In the process of data identification, it is essential to identify the subsets available in the data. We cannot get any valuable information from raw data. Meaningful message that is being retrieved from the data is known as information [2]. Data analysis is the set of schematic process that converts the original data into information and business value. This process selects processed variable as input and converts into the value of information to data analysts. They are using the posts, likes, shares, geography and demographics. After that they will build the data model. In this data model process, they organize the data and find the interrelation between them. Fig. 1 exhibits the schematic diagram of Social Media Analytics Process.

Social Media Audience Segmentation, Information Discovery, Behaviour Inferences, Impact and Exposure are most commonly used cases for SMDA. These cases will give the business insights in terms of analysing the brand performance, comparative feedback with competitors and relevant issues, etc. Bogdan and Philip [3] (2020) presented comprehensive tools used for analysing social media data. They

K. Anitha (✉)
Department of Mathematics, SRM Institute of Science and Technology, Chennai, India
e-mail: anithak1@srmist.edu.in

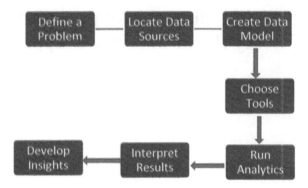

Fig. 1 Social Media Analytic Process

provided the leading software tools used for scrape, cleanse and analyse the social media. Kaplan and Haenlein [4] (2010) defined that the content is ubiquitously accessible. Lazer [5] (2009) stated that it is important for research both qualitative and quantitative techniques. Wolform [6] (2010) took Twitter data to train SVR model to predict individual stock prices of NASDAQ in 15 min earlier. Salathe´ [7] (2012) developed the model for disease transmission rate with the help of news feed and blogs.

1.1 Terminologies

- **Natural Language Processing (NLP):** NLP produces the interrelations between computers and human natural languages through AI. In this process natural language like speech, text, etc. will be manipulated automatically. This process will fill the gap between human communication and understanding of computer. Understanding human language is a complex task. By using NLP through supervised, unsupervised and deep learning techniques we can make construct the model for human language. It involves many steps likes Lexical, Parsing and Semantic Analytics. Lexical analytics divides the entire text into paragraph, new sentences and cluster of words. Parsing does the grammatical arrangement of words. Semantics analytics identifies the exact meaning of each word and creates the mapping between the objects and target domain. After the completion of above-mentioned process finally the given information is converted in to real world knowledge.
- **Text Analytics:** It processes information retrieval to handle word frequency, annotation, pattern identification, posting and tagging. It translates the unstructured text into valid quantitative data to identify the patterns between the texts. These patterns will enable the manufacturers to make wise decision in the aspect of customer satisfaction. For example, if a customer gives the feedback in the form of free text, the analyzation process takes long time. Now many text analytics software is available. It finds the hidden patterns from the customer's

feedback which will create the closed loop between the companies and the end users. Word spotting and text categorization are important steps involved in text analytics. Confirmit, RapidMiner, IBM Watson Studio are well-known software used for text analytics.

- **Sentiment Analysis:** It refers NLP, text analysis to extract the knowledge from source material. It is the process of collecting the categorical opinion about some object through computer. The object may be particular product, topic, etc. Most of the companies will do the process systematically to collect the feedback about their product. It is very much useful if we collect the customers' feedback in voice format or voice of employee feedback. Bain & co stated its research statement that good experiences can grow up to 8% of revenue by increasing the life cycle of customer up to 14%. Stanford NLP, Rosette, Social Mention are some of the well-known sentiment analysis tools.
- **Text Cleaning:** Google Refine and Data Wrangler are some of the text cleansing tools available in online. To remove the noisy data they are using a transformation process by converting the text into meaningful form. This is the pre-processing stage in Sentiment Analysis. Before starting the review removal of emojis, punctuation can be removed at this stage. Normalization, Stop Words, Stemming are some methods used in text cleaning. Normalization removes the noisy information from the text like punctuation, numbers and hashtags. It is like pre-processing stage. Stop words commonly used to text classification process. For example, the word "Not" to be removed from the task box. If it removes the word "Not" from "I am not satisfied", it will give the reverse meaning. For this type of cases stop word can be used. NLTK-Python library has several built-in techniques to eliminate stop words.
- **Opinion Mining:** In this process we can extract the opinion of employees, customers and consumers, etc. for further process about the project, product or news. Automatically it will collect the opinion line positive, negative or neutral opinion, etc. Even it can analyse the personal opinion also. Po-Wei Liang, Bi-Ru Dai (2013) used microblogging of Twitter data.

1.2 Social Data Analysis—Mathematical Perspective

The relational information is the notable feature in social media data. It should concern with uncovered patterns in connection between entities. Data analysts use two kinds of mathematical tools to identify the pattern between the social media data: Graph Networks and Set Theoretical Approach. The most general form of matrix is binary matrix.

$$B_{ij} = \begin{cases} 1 & \text{For Tie}(i \neq j) \\ 0 & \text{There is no tie}(i \neq j) \end{cases}$$

Here social space is mapped by the relation. If $i = j$, they are main diagonal elements. Sometimes it may be important. The purpose of relational database from social media emphasizes the fact that each and every individual has a connection with other individual. Mathematically it may be stated that every object in the universe of discourse is connected with at least one of the other objects from the universal set. Social media data analysis is encountered in finding the patterns (inter relations) between social units. The dependency among the features is measured with different variables. Data analysts confirmed that these social units are not acting independently but they influence each other. Hence finding patterns from these units and identifying the most influential units are essential task for them. But in the approach of set theory, it analyses the data in two ways both qualitative and quantitative. This approach focusses multi-dimensional association and fit the best model. Set theoretical approach periodically constructs the logical and statistical models using continuous variables. Many mathematical tools are available based on set theory to perform this task. Out of these several methods this chapter proposes the approach from the theory of rough sets.

The concept of rough graph is implemented in this paper by the construction of set approximation for rough graph. Here each edge $e \in E$, where E – the edge set of rough graph (G, E) is given by the weight which is defined by the following expression

$$\text{Weight}_i(e_i) = \frac{S(B_{\alpha(i)}, \beta)}{\sum_{i=1}^{n} S(B_{\alpha(i)}, \beta)}$$

where each $i = 1, 2, \ldots n$

$$S(B_{\alpha(i)}, \beta) = 1 - \frac{|B_{\alpha(i)} - \beta|}{\sum_{i=1}^{n} |B_{\alpha(i)} - \beta|}$$

$$\beta = \text{Mean of } B(\alpha_i)$$

1.2.1 Structure of the Chapter

Section 2 describes the foundations of rough sets and its essential terminologies like Reduct and Core. Section 3 exhibits Data Analysis Techniques of Rough Set Theory. Section 4 introduces rough graph along with properties and how reducts are being calculated using graph approximation. Section 5 brings experimental results and Sect. 6 exhibits conclusion along with future work.

2 Rough Sets: Foundations

It is a mathematical approach to handle the data with uncertainty and vagueness. It was introduced by Polish Mathematician Pawlak in 1982 with his representative publication in rough sets and Theoretical Reasoning about data [8, 9]. A rough set is defined by pair of crisp sets called upper and lower approximation which are derived from vague or uncertain information [10]. Elements which are completely belong to the concepts form lower approximation, whereas upper approximation is possibly belong to the concept and their difference is known as boundary region. If boundary is empty set, then the given concept is crisp otherwise it is rough [11]. Equivalence and similarity relations are basics of rough sets in which objects are being classified with same information.

2.1 Terminologies of Rough Sets

Information System: This is the data representation system in rough sets which creates data table in which objects are displayed in rows and their attribute values are displayed in columns. Features or attributes are also known as variables or feature [12] (Polkowski 2002).

An information system $I_s = (\mho, Å)$ is a pair of non-empty set of finite objects ($\mho$) and non-empty finite set of attributes (Å) such that a:$\mho \rightarrow \Delta_a$ for every a ϵ Å, where Δ_a is attribute value set of a. Example for information system is exhibited as follows (Table 1) with 5 objects and 3 conditional variables.

Decision System(D_S): If we include the decision variable column to the information system then such system is known as decision system which will give the decision for each object (Pawlak, 1982). The information system is consistent if and only if for every finite set of objects have same values for their attributes (Rauszer and Skowron, 1992). Table 2 brings the decision system for the above information system.

Indiscernibility Relation: For any $P \subseteq Å$ there will be an equivalence relation IND_P defined as

$$IND_P = \{ (x_1, x_2) \, \epsilon \mho / \forall a \, \epsilon \, P, a(x_1) = a(x_2) \} \tag{1}$$

Equation (1) means that x_1, x_2 are equivalent if their attribute values are same.

Table 1 I_s-Data Table

Objects(O_i)	$ATTR_A$	$ATTR_B$	$ATTR_C$
O_1	a_1	b_1	c_1
O_2	a_2	b_2	c_2
O_3	a_3	b_3	c_3
O_4	a_4	b_4	c_4
O_5	a_5	b_5	c_5

Table 2 Decision System (D_S)

Objects(O_i)	Conditional features			Decision features
	ATT_A	ATT_B	ATT_C	ATT_D
O_1	a_1	b_1	c_1	d_1
O_2	a_2	b_2	c_2	d_2
O_3	a_3	b_3	c_3	d_3
O_4	a_4	b_4	c_4	d_4
O_5	a_5	b_5	c_5	d_5

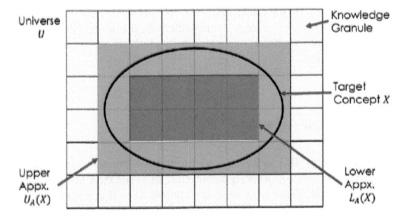

Fig. 2 Rough set—Diagrammatic Form

Set Approximations: Let $X \subseteq \mho$, the approximations ($\underline{P}X$, $\overline{P}X$) are defined as

$$\underline{P}X = \{x / [x]P \subseteq X\} \quad (2)$$

$$\overline{P}X = \{x /_{[x]P} \bigcap X \neq \emptyset\} \quad (3)$$

Fig. 2 gives the details of Rough Approximation

Each and every decision classes from $\{D_1, D_2, \ldots, D_k\}$ are approximated by its pair of approximation on the set $P \subseteq A$ of attributes and they will provide the description of ($\underline{P}X_i, \overline{P}X_i$).

Let P and Q be set of all attributes creating the positive, negative and boundary regions through the equivalence relations over $\mho$.

$$POS_{P(Q)} = \bigcup_{X \epsilon \mho/Q} \underline{P}X \quad (4)$$

$$NEG_{P(Q)} = \mho - \bigcup_{X \epsilon \mho/Q} \bar{P}X \quad (5)$$

$$\text{BND}_{P(Q)} = \bigcup_{X \epsilon \mho/Q} \bar{P}X - \bigcup_{X \epsilon \mho/Q} \underline{P}X \qquad (6)$$

Dependency of Attributes (ϑ)
For $P, Q \subset \text{Å}$, the dependency of Q on P is defined by

$$\vartheta = \beta_P(Q) = \frac{|\text{POS}_{P(Q)}|}{|\mho|} \qquad (7)$$

where $|\mho|$ denotes number of elements in $\mho$.

If $\mho = 1$, Q completely depends on P, if $0 < \vartheta < 1$, there is a partial dependency between P and Q, if $\vartheta = 0$, then P & Q are independent. Based on the dependency value we can conclude that whether the feature should be significant or not. If $\vartheta = 0$ for any, then the feature is inconsistent and removal of this feature will not affect the quality of data [12]. The relation between significant is given by

$$\mu_P(Q, a) = \beta_P(Q) - \beta_{P-\{a\}}(Q) \qquad (8)$$

Reduct: It is the concise or minimal representation of given data set which will describe all information about the original data [9]. The minimal representation of R is defined by

$$\beta_R(D) = \beta_c(D) \qquad (9)$$

where C the set of initial is attributes and D is a given set of attributes.

R is minimum when (2.10) holds

$$\beta_{R-\{a\}}(D) \neq \beta_c(D) \text{ for all } a \epsilon R \qquad (10)$$

The above expression indicated that removal of the subset will definitely affect the quality of data. Then n number of possible reduct sets are given by

$$R_{\text{ALL}} = \{X/X \subseteq C, \beta_R(D) = \beta_c(D) \quad \beta_{R-\{a\}}(D) \neq \beta_c(D) \text{ for all } a \epsilon X \qquad (11)$$

The intersection of all possible reduct sets in R_{ALL} is called Core. $R_{\text{MIN}} \subseteq R_{\text{ALL}}$ is the condition for identifying a single element in the reduct set.

Discernibility Matrix: Calculation of dependency of attributes is essential task in data analysis techniques. One of the effective approaches in RST is discernibility matrix and it was introduced by Skowron and Rauszer in 1992. Each entry in the matrix denotes the set of attributes which are discerns with the corresponding pair.

The discernibility matrix (M) is a for the decision system ($U, C \cup D$) is defined by

Table 3 Inf. Sys(I_S)

Universal set U	a_1	a_2	a_3	a_4	a_5	Decision attribute
o_1	Max	Inter	Inter	F1	P1	Max
o_2	Inter	Inter	Inter	F2	P1	Max
o_3	Inter	Max	Very low	F1	P2	Max
o_4	Inter	Max	Inter	F1	P1	Min
o_5	Min	Inter	Very max	F1	P1	Min
o_6	Max	Inter	Inter	F3	P2	Min
o_7	Max	Max	Inter	F1	P2	Min

Table 4 Discernibility Matrix for Table 3

–	–	–	–	–	–
(a_1, a_4)	–	–	–	–	–
(a_1, a_2, a_3, a_5)	(a_2, a_3, a_4, a_5)	–	–	–	–
(a_1, a_2)	(a_2, a_4)	(a_3, a_5)	–	–	–
(a_1, a_3)	(a_1, a_3, a_4)	(a_1, a_2, a_3, a_5)	(a_1, a_2, a_3)	–	–
(a_4, a_5)	(a_1, a_4, a_5)	(a_1, a_2, a_3, a_4)	(a_1, a_2, a_4, a_5)	(a_1, a_3, a_4, a_5)	–
(a_2, a_5)	(a_1, a_2, a_4, a_5)	(a_1, a_3)	(a_1, a_5)	(a_1, a_2, a_3, a_5)	(a_2, a_4)

$$c_{ij} = \{ a \in C \; ; a(x_i) \neq a(x_j) \}, i,j = 1, 2, .. |U| \} \tag{12}$$

Each c_{ij} contains the attributes with distinct attribute values.

The matrix M symmetric since $M_{(x_i, x_j)} = M_{(x_j, x_i)}$ and $M_{(x_i, x_i)} = \varnothing$ then it is enough to consider upper or lower triangular matrix.

The following table represents the information system (Table 3)

The discernibility matrix for the above table is defined by (Table 4)

This matrix M only considers those object discernibilities that occur when the corresponding decision attributes differ which is useful for finding reducts [12]. Once discernibility matrix M is found, the *discernibility function* f_A for the information system A, a Boolean function f_A of k Boolean variables $a_1^*, a_2^*, \ldots a_k^*$ is defined as

$$f_A(a_1^*, a_2^*, \ldots a_k^*) : \bigwedge \{ \bigvee c_{ij}^* : 1 \leq j, c_{ij} \neq \varnothing \} \tag{13}$$

where $c_{ij}^* = \{ a^* : a \in c_{ij} \}$.

From the above equation it is observed that $\{a_1, a_2, a_5\}$ and $\{a_1, a_2, a_3, a_4\}$ are two reducts for the above information system. By using Eq. (13) it is possible to find the minimum discernibility matrix which are defined as follows (Tables 5 and 6)

Table 5 Minimum Discernibility Matrix {a_1, a_2, a_5}

–	–	–	–	–	–
a_1	–	–	–	–	–
a_1	a_2	–	–	–	–
a_1	a_2	a_5	–	–	–
a_1	a_1	a_1	a_1	–	–
a_5	a_1	a_1	a_1	a_1	–
a_5	a_1	a_1	a_1	a_1	a_2

Table 6 Minimum Discernibility Matrix {a_1, a_2, a_3, a_4}

–	–	–	–	–	–
a_4	–	–	–	–	–
a_2	a_2	–	–	–	–
a_2	a_2	a_3	–	–	–
a_3	a_3	a_2	a_2	–	–
a_4	a_4	a_2	a_2	a_3	–
a_2	a_2	a_3	a_1	a_2	a_2

3 Rough Sets: Data Analysis Techniques

Rough set attribute reduction process identifies equivalence classes between the attributes. A minimal representation of the universal set which is known as reduct that preserves the property of partitioning of the whole set and which is having the ability to create the same classification as the entire dataset.

Let R be the set of all reducts then the minimal reducts $R_{Min} \subseteq R$ can be defined as

$$R_{Min} = \{ Y : Y \in R, \forall T \in R, |Y| \leq |T| \} \tag{14}$$

Elements of core cannot be eliminated from the given set.

3.1 Rule Discovery on Rough Set Theory

Rule discovery is an essential task since the relationships in data are in the form of "If C then D" do not necessarily reflect real rules of the application domain, and other problems. Therefore, there is a need to eliminate incorrect rules. The construction of a rule discovery algorithm with estimated error rates of classification could be developed by the rough sets theory which is illustrated by the following example. Table 3.1 be the decision system with four conditional attributes(c_1, c_2, c_3, c_4) and a decision attribute (d) from which rules are generated and they are expressed in Tables 7 and 8.

Hu (2004) proposed new algorithm which embedded relational algebra in rough sets. Let C be a set of conditional attributes and D-Decision attribute then the reduct A

Table 7 Decision Table

c_1	c_2	c_3	c_4	d
0	0	0	1	0
0	0	1	3	0
0	1	0	2	0
0	1	1	0	1
1	1	0	2	2

Table 8 Set of Rules for Decision Table

If $(c_2 = 0)$ then $d = 0$
If $(c_1 = 0)$ & $((c_4 = 2)$ then $d = 0$
If $(c_4 = 0)$ then $d = 1$
If $(c_1 = 1)$ then $d = 2$

is defined to be a subset $RED_A \subseteq C$ with the decision attribute D. RED_A is a minimal subset of attributes which has the same classification power as the whole condition attributes.

Let K be the degree of dependency between RED_A and D, K' be the probability of data occurrence in the decision table, *card* -denotes count operation in data bases, Π- projection operation in data bases then stopping criteria of algorithm are defined as

$$K' = \frac{\text{Card}(\Pi(RED_A + D))}{\text{Card}(\Pi(C + D))} \tag{15}$$

The impact of each conditional attribute on the decision attribute D is called measure of merit which is defined by

$$\text{Merit}(C_i, C, D) = 1 - \frac{\text{Card}(\Pi(C - \{C_i\} + D))}{\text{Card}(\Pi(C + D))} \tag{16}$$

4 Rough Graph

This chapter introduces rough set theory based deterministic rules in terms of graphical structure. Each member in the Facebook is considered as vertex (nodes) and their friendship relations are denoted as edges (links). The theory of rough graph is an attempt to merging the properties of classical graph in rough sets. The amount of data in Facebook is rapidly increasing in exponential way. Finding the pattern among the relations with attributes such as likes, dislikes, comments, etc. will make good impacts in decision making. When we are finding ambiguity in between relation it is very difficult to fix the rule. These types of issues can be solved by the construction of rough graphs.

He Tong [13] introduced the structures of rough graph in 2006. They extended their work in 2019 as Vertex Rough Graph. It is the pair of two crisp graphs called lower and upper graphs. Let $G = (V, E)$ be the graph with vertex set V and edge set E. Let R be the equivalence relation defined on V and $[v]_R$ is the equivalence classes on v. Let $S(V', E')$ be R - Vertex Rough Graph of G if it satisfies the following conditions:

$$R_*(V') = \{v \in V : [v]_R \subseteq V'\}$$

$$R^*(V') = \{v \in V : [v]_R \cap V' \neq \emptyset\}$$

$$R_*(E') = \{(v_r, v_t) \in E', v_r, v_t \in [v]_R \text{ for some } v \in R_*(V') = \{v \in V : [v]_R \subseteq V'\}$$

$$R^*(E') = \{(v_r, v_t) \in E', v_i \in [v_r]_R, [v_i]_R \bigcap E' \neq \emptyset \text{ and } v_j \in [v_t]_R, [v_t]_R \bigcap E' \neq \emptyset$$

The pair $(R_*(S), R^*(S))$ is called R- Vertex Rough Graph.

4.1 Discernibility Matrix for Rough Graph

Let $S(V, E)$ be the graph and $S(V', E')$ be the rough graph. Then the following expression represents the discernibility matrix

$$DM_{ij} = \begin{cases} 0 & \text{if } (v_r, v_t) \notin R^*(V') \\ 1 & \text{if } (v_r, v_t) \in R^*(V') \text{ and} \notin R_*(V') \\ 2 & \text{if } (v_r, v_t) \in R_*(V') \end{cases}$$

This paper finds the pattern based on friendship relation between Facebook contacts by using the concepts of Vertex Rough Graph. Let $S = (V, E)$, where $V = \{f_1, f_2, f_3, f_4, f_5\}$, $\forall f_i$ denotes the person and their relations are described by the following graph (Fig. 3).

Graph $S(V', E')$ where $V' \subseteq V$, $E' \subseteq E$ is defined by (Fig. 4).

$$R_*(V') = \{(f_5, f_2, f_4, f_3), (f_1)\}$$

$$R^*(V') = \{f_1, f_2, f_3, f_4, f_5\}$$

$$R_*(E') = \{(e_5, e_{10}, e_8, e_9, e_4), (e_6, e_7)\}$$

$$R^*(E') = \{e_5, e_{10}, e_8, e_9, e_4, e_6, e_7, e_1, e_2, e_6, e_3\}$$

$$R^*(S) = (R^*(V'), R^*(E')), R_*(S) = (R_*(V'), R_*(E)').$$

The pair $(R_*(S), R^*(S))$ is Vertex Rough Graph. The discernibility matrix for this example is given below:

Fig. 3 Graph S (V, E)

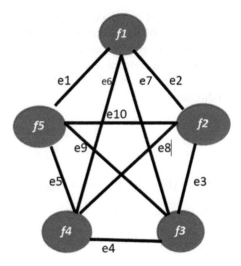

Fig. 4 Graph S (V′, E′)

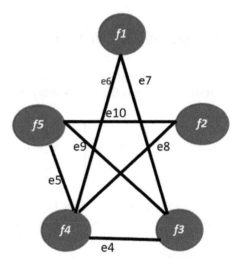

$$DM_{ij} = \begin{pmatrix} 0 & 0 & 2 & 2 & 1 \\ 0 & 0 & 1 & 2 & 2 \\ 2 & 1 & 0 & 2 & 2 \\ 2 & 2 & 2 & 0 & 2 \\ 1 & 2 & 2 & 2 & 0 \end{pmatrix}$$

If $f_i \rightarrow n$ the graph can be drawn in the following way (Fig. 5).

This paper uses similarity relation between nodes (Vertices) instead of taking indiscernibility relation which is described by

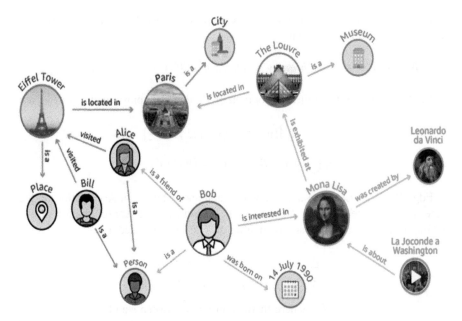

Fig. 5 Facebook network graph

Fig. 6 Graph Similarity Relation

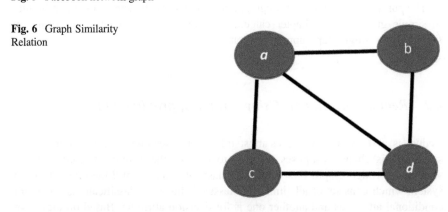

$S = \{[x]$ such that $x, y \in U,$ where xRy and yRx $\}$, where $xRy - x$ related to y. David Nagy [14] designed the graphical structure based on similarity relation as there exists a positive edge between the objects if they are similar otherwise the edge is negative.

The following set describes the connectivity between the nodes based on similarity relation of Graph G (Fig. 6)

$$S = \{(a,b,+), (b,a,+), (b,d,+), (a,d,-), (d,b,+), (c,d,+), (d,c,+), (a,c,+), (c,a,+)\}$$

For attribute reduction the base vertex graph is framed by using this similarity relation. Base graph B is given by

$$B_*(V) = \{v \in V_1 \text{ such that } v \text{ is related to } u \text{ iff } v(x) = v(y) \forall x, y \in V_1\}$$

Let $V_2 \subset V_1$ be the arbitrary subgraph of V_1, then set approximations of V_2 are

$$\underline{V_2} = \bigcup \{v; v \in B_*(V) \text{ and } v \in V_2\}$$
$$\bar{V_2} = \bigcup \{v; v \in B_*(V) \text{ and } B_*(V) \cap V_2 \neq \emptyset\}$$

Edge R-graph: Let K be the complete graph and A be the binary relation on the edge set of K. Then the pair $E = (K, A)^c$ is known as Edge Approximation on the R-Graph. Lower and upper approximation of $X \subset E$ is defined as

$$\underline{A}(X) = \{x \in E(K), E_A(x) \subseteq X\}$$
$$\overline{A}(X) = \{x \in E(K), E_A(x) \cap X \neq \emptyset\}$$
$$E_A(x) = \{e_i; \text{ where there is an edge between } e \text{ and } x\}$$

The purpose of existence of rough graph is to identify the vagueness or ambiguity in simple graphs. This chapter calculate the reduct from the given information system using rough graph approximations.

4.2 Reduct Calculation Using Graph Approximation

Reduct is a set which preserves most informative features of given information system. This chapter proposes the idea to remove the redundant feature by the structure of rough graph. First similarity class table has to find the given decision system which consists of all similarity classes. It has two classifications one is for conditional attributes and another one is for decision attribute. Based on these two classifications it may be extended for boundary values and positive region. After these constructional processes rough graph is constructed based on discernibility relation. Each entry from the discernibility matrix is represented by a vertex(v_i) for this matrix. Add all the vertices(v_j) one by one if the following conditions are hold

$$V_{add} = \{v_j \text{ iff } a(v_i) \neq a(v_j), v_j \in \text{postive region}\}$$

In this process high level of accurate variable will only be selected and then this process is framed as decision rules. The following diagram (Fig. 7) exhibits rough set theory based data analysis process.

The schematic process of rough set data analysis is explained below

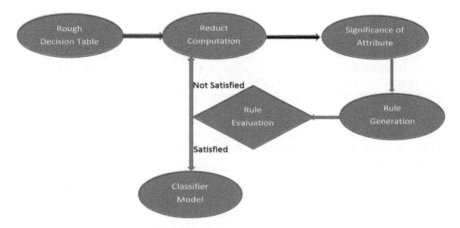

Fig. 7 Rough Set Data Analysis

- Create decision table—Based on discernibility matrix.
- Construction of similarity classes.
- Create the rough vertex graph based on similarity relation.
- Construction of rough discernibility function.
- Rule generation—Rules will be generated with reference to the network relation.
- Rule evaluation.
- Compute reduct with minimal number of attributes.
- Calculate the significance of attributes.

5 Experimental Analysis

Data Set: Facebook Pseudo data set is taken from Kaggle. This data set contains 99,903 objects with 15 attributes. This exploratory data analysis is helpful to identify the users that can be focused more on to increase the business.

Variables: User id(a_1), Age(a_2), DoB_DaY(a_3), DoB_Year(a_4), DoB_Month(a_5), gender(a_6), tenure(a_6), Number of Friends(a_7), friendship initiated(a_7), likes received(a_8), mobile likes(a_9), mobile-like received(a_{10}), www likes(a_{11}), www likes received(a_{12}).

These 12 attributes are considered as edges (links) and number of persons are taken as vertices. In this process the strongest link has been identified to develop the network.

Tool: Rosetta.

Rough Set Discernibility Function Is Being Created for Entire Data Set

```
f(0) = (User_ID + Dob-Month + tenure) *(User_ID + Age$ + DoB-Year +
tenure) *(User_ID + Dob-Day + dob_month) *(User_ID + Age$ + Dob-Day +
dob_year) * (User_ID + Dob-Day + tenure)
```

end
f(1) = (User_ID + Age$ + Dob-Day + DoB-Year + gender + likes + Likes_Received + Mobile-Likes + mobile Likes_Received + www.likes + www..likes..received) * (User_ID + Age$ + DoB-Year + Dob-Month + gender) * (User_ID + Dob-Day + Dob-Month + gender) * (User_ID + Age$ + Dob-Day + DoB-Year + dob_month) * (User_ID + Age$ + DoB-Year + tenure) * (User_ID + Dob-Month + tenure) * (User_ID + Dob-Day + gender + tenure)
end
f(2) =
 (User_ID + Dob-Month + tenure) * (User_ID + Age$ + DoB-Year + dob_month) * (User_ID + Dob-Day + dob_month) * (User_ID + Age$ + DoB-Year + tenure) * (User_ID + Dob-Day + tenure)
end
....
....

f(3422) = (User_ID + Dob-Day + Dob-Month + tenure + friend...count + friendships_initiated + likes + Mobile-Likes) * (User_ID + Age$ + Dob-Day + DoB-Year + tenure + likes + Likes_Received + Mobile-Likes + www.likes + www..likes..received) * (User_ID + Dob-Day + tenure + Likes_Received + Mobile-Likes + mobile Likes_Received + www_likes) * (User_ID + Dob-Day + Dob-Month + likes + Likes_Received + Mobile-Likes + Received Mobile Likes) * (User_ID + Dob-Day + Dob-Month + tenure + friend...count + likes + Likes_Received + Mobile-Likes + www.likes + www..likes..received) * (User_ID + Age$ + DoB-Year + Dob-Month + tenure + likes + Mobile-Likes + mobile Likes_Received + www.likes + www..likes..received) *
 (User_ID + Dob-Day + Dob-Month + gender + tenure + likes + Likes_Received + mobile Likes_Received + www_likes) * (User_ID + Dob-Day + Dob-Month + tenure + friend...count + friendships initiated + Likes_Received + Received Mobile Likes) * (User ID + Dob-Day + gender + tenure + friend... count + friendships_initiated + likes + Likes_Received + mobile Likes_Received + www_likes) * (User_ID + Age$ + Dob-Day + DoB-Year + Dob-Month + tenure) * (User_ID + Age$ + Dob-Day + DoB-Year + Dob-Month + Likes_Received + Received Mobile Likes) * (User_ID + Dob-Day + Dob-Month + tenure + likes + Mobile-Likes + mobile Likes_Received + www..likes..received) * (User_ID + Dob-Day + Dob-Month + gender + tenure + friend... count + friendships_initiated) * (User_ID + Age$ + Dob-Day + DoB-Year + gender + tenure + likes + Mobile-Likes) * (User_ID + Age$ + Dob-Day + DoB-Year + tenure + friend...count + friendships_initiated + Likes_Received + www..likes..received) * (User_ID + Age$ + Dob-Day + DoB-Year + tenure + friend...count + likes + Mobile-Likes + mobile Likes_Received + www..likes..received) * (User_ID + Age$ + DoB-Year + Dob-Month + gender + tenure + Likes_Received + Received Mobile Likes) * (User_ID + Age$ + Dob-Day + DoB-Year + tenure + Likes_Received + Received Mobile Likes) * (User_ID + tenure + likes + Likes_Received + Mobile-Likes + Received Mobile Likes) * (User_ID + Dob-Day + Dob-Month + gender + tenure + friendships_initiated + likes + Mobile-Likes) * (User_ID + Age$ + DoB-Year + tenure + friend...count + friendships_initiated + Likes_Received + Received Mobile Likes) *

From the above discernibility function 6548923 non-deterministic rules have been generated for the data set. Based on the decision rules the following reducts have been found for the given data set.

S. no	Reduct	Support	Length
1.	{User ID}	100	1
2.	{age, tenure, likes_received}	100	3
3.	{likes_received, Mobile_likes received}	100	3
4.	{dob_Year, likes_received, Mobile_likes received}	100	3
5.	{friendship initiated, likes_received, Mobile_likes received}	100	3
6.	{age, likes_received, Mobile_likes received}	100	3
7.	{dob_Year, likes_received, tenure}	100	3
8.	{dob_day, likes_received, mobile likes received}	100	4

From the above result user id, Age$, tenure, like received, mobile likes received, www likes, friendship initiated are some of the reducts with the support of 100. They are the possible reduct sets. Intersection of all reducts will give the value of core. Here mobile likes received is the most important attribute and it will provide the strongest network. Hence mobile likes received is the key attribute and removal of this variable completely changes the data. These valuable insights will give the suggestions to Facebook to make wise decision to identify its valuable users and provide the novel recommendations to them.

6 Conclusions and Future Work

This paper concludes that rough set analyses the pattern only hidden in the data by using similarity relation through its rough graph. Relation between the objects can be identified by their similarity relation. It handles the data based on approximation space which will enable the data scientists to prioritize the boundary values which depends on pair of approximations. In this paper data analyzation was done through rough graph. The approximation has been found by the relationship between the vertices and it is implemented in Facebook data set. The extended part of this concept can be in rough hybridization with near set which will consider the objects based on neighbourhood System. Rough hybridization with near set will give the optimal result in image filtering.

References

1. Marshall, S., & Khan Gohar, F. (2017). *Digital analytics for marketing*. Routledge. ISBN 9781138190672. OCLC 975370877.
2. Singh, S., Arya, P., Patel, A., & Tiwari, A. K. (2019). Social Media Analysis through Big Data Analytics: *A Survey, Proceedings of 2nd International Conference on Advanced Computing and Software Engineering* (ICACSE).
3. Batrinca, B., & Treleaven, P. C. (2015). Social media analytics: A survey of techniques, tools and platforms. *Open Forum, AI & Soc, 30*, 89–116. https://doi.org/10.1007/s00146-014-0549-4

4. Kaplan, A. M., & Haenlein, M. (2010). Users of the world, unite! The challenges and opportunities of social media. *Business Horizons, 53*(1), 59–68.
5. Lazer, D. (2009). Computational social science. *Social Science, 323*, 721–723.
6. Wolfram SMA (2010) Modelling the stock market using Twitter. *Dissertation Master of Science thesis, School of Informatics.*
7. Salathe, M. (2012). Digital epidemiology. *PLoS Computational Biology, 8*(7), 1–5.
8. Nayyar, A., & Puri, V. (2017). Comprehensive analysis & performance comparison of clustering algorithms for big data. *Review of Computer Engineering Research, 4*(2), 54–80.
9. Pawlak, Z. (1982). Rough sets. *International Journal of Computer and Information Sciences, 11*, 341–356.
10. Kumari, A., Behera, R. K., Sahoo, K. S., Nayyar, A., Kumar Luhach, A., & Prakash Sahoo, S. (2020). Supervised link prediction using structured-based feature extraction in social network. *Concurrency and Computation: Practice and Experience, 58*(39), –e5839.
11. Abraham, A., Falcon, R., & Bello, R. (1999). *Rough set theory: A true landmark in data analysis - studies in computational intelligence* (p. 174). Springer.
12. Polkowski. (2002). *Rough Sets, Mathematical foundation.* Springer-Verlag.
13. He, T., & Shi, K. (2006). Rough graphs and its structure. *Journal of Shandong University (Nature Science), 6*, 88–92.
14. Dávid, N., Tamás, M., & László, A. (2020). Graph approximation on similarity based rough Sets, an international journal for engineering and information sciences. *Pollack Periodica, 15*(2), 25–36.

Index

A
Active Antenna Unit (AAU), 117
Adam optimizer, 148, 150, 152, 159
Advanced Research Projects Agency (ARPA), 102
Agribots, 90
Agricultural industry, 94
Agriculture, 92
AI advantages in education sector
  easy curriculum development, 192
  global connect among peers, 194
  learning anytime anywhere, 192
  online assessment, 192
  personalized learning experience, 193
  real-time feedback, 194
  student engagement, 192, 193
  student facilitation, 192
  student's performance improvement, 193
  subject expert supported learning, 192
  time-saving, instructors, 192, 193
  virtual mentoring, 192
AI-based educational technologies, 190
AI-based educational solutions, 196
AI-based education system, 206
AI-based technologies for education
  personalized learning, 194
  smart digital content, 195
  voice assistance, 195
  world-class learning experience, 195
AI-driven educational platforms, 205
AI-driven system, 190
AI-enabled applications/systems, education sector
  academic management personnel, 190
  advantages (*see* AI advantages in education sector)
  automated ERP management systems, 190
  data security, 190
  educational solutions, 195
  e-learning, 190
  global and personalized learning, 189
  implementation criteria, 194–195
  SIS, 190
  smarter content creation, 189
  technologies, 190
AI-enabled online learning platform development
  content repository creation, 195
  know your competitors, 195
  know your project requirements, 196
  overview, 197
  regular system updation, 197
  solution launch and customer feedback, 197
  testing and debugging, 196
AI innovation, 190
Analyzation process, 210
Artificial cyber espionage
  automated plans, 172
  and deep learning algorithms, 170
  search engines, 168
  technologies, 169
Artificial intelligence (AI), 40, 78, 168
  agriculture, 88
  agriculture sector, 90
  applications, 90
  crop, 91
  data, 84
  definition

Artificial intelligence (AI) (cont.)
    Coppin, 189
    Whitby, 189
  human work, 84
  role, 90
  scope, 91
  techniques and algorithms, 84
Artificial intelligence-internet of things (AIIoT)
  advantages, 87
  AI and ML methods, 87
  algorithms, 79
  analytical techniques, 79
  automatic alarms, 87
  automatic vacuum cleaner, 87
  disadvantages, 79
  DTAC, 95, 96
  GPS machinery, 96
  network system, 79
  Vodafone company, 96
Artificial neural networks (ANNs), 43
Attribute dependency, 215
Automated AI, 169
Automated ERP management systems, 190
Automated metropolis surveillance, 172, 173
Autoradiopuhelin (ARP) technology, 106

B
Base station (BTS), 104, 105
Base station controller (BSC), 105
Beam forming, 120, 121
Body sensor network (BSN), 62

C
Cameras, 181
Carnegie Learning, 194
Center for Disease Control (CDC), 56
Cloud computing, 13
Cloud-IoMT-based architecture
  computer and AI data analysis model, 65
  gateway, 66
  healthcare experts, 65
  isolation/quarantine layer, 65
  layers, 63
  storage database, 65, 66
  symptoms layer, 64
Cloud Vision API, 204
Convolutional neural networks (CNNs), 147, 148
COVID-19 pandemic
  business fields, 61
  capacity, 61
  cloud-based IoT administrations, 58
  cloud computing, 60
  CloudIoT-Health, 61
  confusion matrix, 68, 69
  demands, 57
  designing smart applications, 56
  diagnosis models, 67
  diagnostic and healthcare devices, 56
  effective techniques, 59
  emerging policies, 60
  features, 61
  healthcare management and monitoring scheme, 62
  healthcare sectors, 56, 59
  healthcare systems, 59
  heterogeneity problem, 62
  home-based diagnostic test, 56
  human resources, 58
  implementations, 56
  infectious disease epidemiology, 58
  integration, 60
  IoT, 60, 63
  layers, 62
  lockdown measures, 55
  medical care, 58
  medical environment, 62
  medical organization, 58
  method, 59
  metrics, 67
  ML, 68
  NS2 and NetSim Simulators, 61
  online health consultations, 56
  parameters, 67
  powerful and supportive medical system, 57
  preprocessing, 66
  public clouds, 60
  ROC curves, 70, 71
  safety measures, 62
  SARS-like epidemic, 58
  sensors, 58
  smart healthcare hospital, 61
  techniques, 62
  telemedicine systems, 59
  tools, 58
  wearable body sensor network, 56
  wearable devices, 62
  wearable healthcare devices, 60
  web-based monitoring platforms and strategies, 60
  web communications systems, 60
  wireless sensors, 56
  World Health Organization, 55
Cyber espionage, 168

Cyber illicit filtrations
    dossier and identity thievery, 178–179
    network information flooding threat, 179
        firmware integrity and impregnable boot, 179
        mutual testament, 180
        protocol engines, 180
        provisioning and key governance, 180, 181
        root of credible, 180
        surveillance contemplating and analysis, 180
        surveillance lifecycle governance, 180
        technological enabled colonies, 179
    perpetrator eavesdrop threat, 177–178
Cybernated-actual articles, 181
Cybernated-physical webnet, 181
Cybernated wrongdoer, 172, 175, 179
Cyber Security Threat Vectors, 175
Cyber Threats Intelligence framework, 174

D

Dark web (DW), 168, 169
Data acquisition systems, 81
Data analysis, 209
Data augmentation, 157
Data identification, 209
Decision rules, 222
Decision system, 213, 214
Decision table, 218
DeepLab V3 model, 149
Deep-learning (DL), 147, 148, 150, 157
Delivery delay time, 47, 49
Delivery Wait Period (DWP), 46
Digital security framework, 174
Digital technologies, 189
Direct line of sight (LOS), 116
Discernibility matrix, 215, 216
DL processor, 18
DNA testing, 178
DSTL Satellite Imagery Feature Detection database, 155
Ducere Technologies, 204

E

Edge approximation, 222
Edge computing, 14, 15
EEG sensors, 205
Efficient Routing System (ERS), 43
Electromagnetic field, 117

Empty set, 213
Energy efficient, 9
ERS-SDN-IOV proposed system layout, 44
Extra Trees, 59, 67

F

Facebook data set, 225
Farm machinery testing, 93
Fifth-generation (5G) cellular network
    architecture, 126, 128, 130
    basic principles, 128, 130
    beam forming, 120, 121
    communications, 101, 141
    device to device communication (D2D), 121, 122
    digital generation, 107
    edge computing, 137, 138
    full duplex, 125
    heterogeneous networks, 116, 117
    hypothesis, 104
    IIoT, 132
    industry and home, 141
    infrastructure, 132
    Internet of Things (IoT), 102, 128, 130
        networks, 136
        protocols, 136
        public, 141
        technologies, 132
        use cases, 140
    IoE, 131
    LTE, 110, 111
    massive MIMO, 117–120
    millimeter waves, 113–116
    mobile broadband evolution, 106, 107
    mobile networks, 104
    multiple accesses, code division, 107, 108
    network, 102
    NG-RAN, 126, 127
    NOMA, 123, 124
    objectives, 103
    physical and virtual world, 133, 134
    recommendations, 142
    research methodology, 104
    security, 138–140
    structure, IoT, 134, 135
    TCP/IP, 102
    technologies, 102, 112, 113
    telecommunications, 101
    television and telephony, 102
    third generation, 108–110
    transmission, 103

Fingerprinting, 178
Fog computing, 15, 16
Fully convolution network (FCN), 147–149

G
Geographical information system (GIS), 43
Graph similarity relation, 221
Green computing
    agriculture, 11
    applications, 9
    communication process, 9
    decision-making process, 10
    development, 9
    green campus, 10, 11
    health care, 11
    intelligent automobiles, 11
    intelligent houses, 12
    intelligent protection/security, 12
    policy-based, 10
Greenhouse gases, 146

H
Hard-Swish, 150
High-resolution remote sensing satellites, 145
Home webnet, 172
Hybrid Software Defined Networking Geographic Routing Algorithm, 41

I
Indiscernibility relation, 213–215, 220
Industrial Internet of Things (IIoT), 132
Information, 175, 176
Information system, 213, 216
Information technologies and communication (ICTs), 101
Intelligent highways, 82
Intelligent vehicle system, 45
Interactive voice assistance services, 195
International Telecommunication Union (ITU), 112
Internet of Medical Things (IoMT), 56
Internet of Nano Things (IoNT), 2
Internet of Things (IoT), 56
    actuators, 81
    AIIoT system, 86
    ambient notification, 20
    and ML, 170
    application domains, 6
    applications, 4, 6, 82, 89
    architecture, 6, 81, 85

architecture view (*see also* IOT architecture)
artificial intelligence, 85
automated lights, 172
awareness programs, 20
big data, 1
businesses, 80
carbon dioxide, 7
categories, 31
challenges, 2
cloud computing, 1, 13
CO_2 emission, 27
communication, 2, 85
computers, 2
concept, 81
data centre, 16, 82
digital signals, 81
ecosystem, 3
edge computing, 14, 15
edge IT, 82
effective decision-making process, 30
electrical grid, 2, 30
electronic devices, 1
elements, 4–6
energy consumption requirements, 8
energy, 7
error-free communication, 6
Fog computing, 15, 16
gateway, 80
green computing, 2, 8 (*see also* Green computing)
green environment, 3
greenhouse gases, 2, 30
hardware-based, 17
healthcare sector, 83
human beings, 1
human–computer interaction, 80
humidity and temperature, 83
integrated circuits, 18
interfaces, 2
Internet play, 2
issues, 2
livestock, 89
machine to machine communication, 2
manufactures, 8
medical resources, 6
middleware layer, 6
natural resources, 2
network layer, 6
physical devices, 4
platform, 82
power consumption, 27, 30
principles, 9
processor, 18

protocols, 80
public sectors, 21
real-time monitoring, 1
recycling, 21, 22
RFID, 17, 18
role, 88
science and technology, 1
segments, 4
sensors, 19, 81
service providers, 8
smart agriculture system, 30, 82
smart devices, 80
smart health care systems, 6
smart homes, 83
smart metering system, 20
smart parking, 82
soil quality, 89
stages, 81
sustainable environment, 27, 29
techniques, 8
telecommuting, 16
transport systems, 7
virtualization, 16
waste materials, 2
weather-related disasters, 2
web-modeling, 179
IoT applications in education
  institutional infrastructure safety measures, 202
  mathematical formulas, 197
  smart infrastructure, 198–200
  students attendance management, 200
  technological advancement, 198, 202–204
IOT architecture
  administration liturgy, 183
  automated gadget/sensing chips tiers, 182
  gateways and webnet, 182
  IOT conceptual view
    abstraction tiers, 184
    access tiers, 184
    equate tiers, 183–184
    service tiers, 185
ISO standards, 95

J
Jaccard coefficient, 152
Jaccard Index, 157

K
Kaggle platform, 155, 159

L
Learning techniques, 210
Legislation enforcement (LE), 168
Lexical analytics, 210
Light Gradient Boosting Machine (LGBM), 59
LinkNet model, 148, 150, 161
Long term evolution (LTE), 110, 111
Loop simulation testing, 92
Loss function, 158

M
Machine learning (ML), 84, 85, 91, 167, 168, 170, 172, 176, 177, 204
Machine learning model (MLM), 43
Machine Learning Module (MLM), 44
Massive Open Online Courses (MOOCs), 205
MCS software, 40
Measure of merit, 218
Minimum discernibility matrix, 216, 217
Mixed Integer Linear Programming (MILP), 16
Mobile Switching Center (MSC), 105
Mobility assistance, 203
Modern mechanization, 35
Multimedia classrooms, 199, 200

N
Natural Language Processing (NLP), 210
Network function virtualization (NFV) mechanisms, 41
NLTK-Python library, 211
Non-line of sight Direct vision (NLOS), 116
Non-orthogonal multiple access (NOMA), 123, 124
Normalization, 211

O
Online satellite images, 149
Open Flow, 41, 42
Open Street Map (OSM), 45
Operational efficiency, 205
Operational technology (OT), 136
Opinion mining, 211
Organic radical battery (ORB), 24
Orthogonal multiple access techniques (OMA), 123

P
Packet delivery ratio (PDR), 46, 47
  performance analysis, 47
  vs. vehicle densities, 48

Packet loss (PL), 50, 51
  performance analysis, 50
  vs. vehicle speed, 50
Permanent Denial of Assistance (PDoA), 179
Personal computers, 189
Personalized education, 206
Polylactic acid plastic (PLA), 24
Principal component analysis (PCA), 157

Q
Quality of service (QoS), 106

R
Radio frequency (RFID), 17, 18
Random forest, 59, 71
R-CNN model, 149
Rectified linear unit (ReLU), 151
Red Technopreneur Colombia, 96
Reduct, 215, 217
Residual blocks, 147, 151, 152, 154, 161
Robot tractor, 92
Rough approximation, 214
Rough graph
  classical graph, 218
  discernibility matrix, 219, 220
  reduct calculation, 222, 223
  similarity relation, 221, 222
  structures, 219
  vagueness/ambiguity, 222
  Vertex Rough Graph, 219
Rough hybridization, 225
Rough set
  attribute reduction process, 217
  definition, 213
  deterministic rules, 218
  discernibility matrix, 215, 216
  equivalence, 213
  indiscernibility relation, 213–215
  information system, 213
  reducts, 217
  rule discovery, 217, 218
  similarity relations, 213
Rough set theory based data analysis
  approximation space, 225
  attributes, 223
  diagram, 222, 223
  discernibility function, 223, 224
  recommendations, 225
  reduct intersection, 225
  schematic process, 222
Rule discovery algorithm, 217

S
Satellite image analysis, 146
  CNN architecture, 149
  green land detection
    data augmentation, 157
    data transformation, 156
    DSTL dataset, 155, 156
    experimentation platform, 155
    Jaccard Index, 157
    LinkNet, 161
    loss function, 158
    ResNet, 161
    segmentation results, 158, 159, 162
  planet database satellite images, 150
  Satellite WorldView-3, 155
  semantic segmentation, 146–148
  SVM's, 149
  U-NET, 148, 151, 163
Satellite image processing, 146
SDN-based vehicle network (SDVN), 42
Seclusion, 175, 181
Semantic analytics, 210
Semantic segmentation
  CNN based framework, 148
  in computer vision application areas, 145
  FCN, 148
  image labeling, 146
  Jaccard Index, 150
  LinkNet and U-Net, 148
  ML, 146
  object classification, 145
  in RS field, 146
  satellite images using U-Net, 148
  skip-U-NET, 147
  U-NET, 147, 152, 163
Sensing chips, 175, 182–184
Sentiment analysis, 211
Set approximations, 214, 222
Simple Network Management Protocol (SNMP), 102
Simulated intelligence, 167
Simulation parameters, 45
Simulation set-up of the network, 45
Skip-U-NET model
  architecture and working, 153, 154
  semantic segmentation, 152
  skip connection, 155
  system model, 152
Smart Cities Models, 170, 171
Smart Cities Security Network, 181, 182
Smart city scenarios, 40
Smart devices, 204
Smart digital content, 195

Index

Smart gadget, 204
Smart homes, 204
Smart infrastructure
   computer-based intelligence, 198
   computer resource allocation, 199
   digital resource, 200
   multimedia classrooms, 199, 200
   personalization and smoothing, 198
   smart interactive boards, 198
Smartphones
   accessories requirements, 24
   development, 22
   energy-efficient hardware components, 22
   energy-saving smartphone batteries, 24
   green computing plays, 22
   green materials, 24
   hazardous materials, 23
   production, 29
   recycling, 23, 24
   reduce packaging, 24
   selling and emission rate, 26, 28
Smart transportation system, 11
Social data analysis, mathematical perspective
   finding patterns, 212
   Graph Networks, 211
   multi-dimensional association, 212
   rough graph, 212
   set theoretical approach, 211
   social space, 212
Social media data analysis (SMDA)
   optimal decision making, 209
   purpose, 209
   schematic diagram, 209, 210
   software tools, 210
   use cases, 209
Software-defined networking (SDN)
   advantages, 36
   concept, 38
   controller, 44
   literature review, 37
   mechanization, 35
   OS type, 39
   VANET model, 37
   VANET system, 38
Software-Defined Vehicular Network (SDVN)
   architecture, 37, 42
Software testing
   compatibility testing, 94
   functional testing, 93
   performance testing, 94
   regression testing, 94
   usability testing, 94

Spiking Neural Network (SNN), 19
Stopping criteria of algorithm, 218
Storage governance, 185
Student information management system (SIS), 190
SVR model, 210

T
Tech-led intervention, 193
Technological advancement
   autonomy improvement, 204
   innovative designs, 202
   mobility assistance, 203
   reading surroundings, 203, 204
Technological enabled colonies, 169, 170, 172, 173, 175, 176
   illicit filtrations (see Cyber illicit filtration)
Teleconferencing software, 16
Test standards, 94
Text analytics, 210
Text cleaning, 211
Traditional learning approaches, 205

U
U-NET model
   CNN, 151
   DeepLab V3, 149
   hybridized model, 147, 151
   in IOU, 149
   and LinkNet, 148
US Environmental Protection Agency (EPA), 2
User-friendly digital resources, 201
Using artificial cyber espionage, 169, 170

V
Variables/feature, 213
Vehicle monitoring sensors, 11
Vehicular ad-hoc networks (VANETs)
   empowering techniques, 38
   ERS-SDN-IOV approaches, 37
   fog computing, 38
   FSDN, 37
   IMD, 39
   literature comparisons, 41
   ONOS, 39
   positioning and administration problems, 37
   SDN, 37, 38

Vehicular intelligent system, 46
Vertex Rough Graph, 219
Virtual reality (VR), 192

W
Web-based education, 205
Web-Modeling, 180, 183
Webnet, 172, 183, 184

Wideband Code Division Multiple Access (W-CDMA), 108
Wireless communication gateway framework, 171
Wireless networks, 102, 116

X
XGBoost, 59, 67, 68, 71

Printed by Books on Demand, Germany